REFERENCE

THE HISTORICAL
ATLAS
OF THE
EARTH

A VISUAL EXPLORATION
OF THE EARTH'S PHYSICAL PAST

GENERAL EDITORS
ROGER OSBORNE
AND
DONALD TARLING

CONSULTANT EDITOR
STEPHEN JAY GOULD

ADDITIONAL CONTRIBUTIONS BY
G.A.L. JOHNSON
UNIVERSITY OF DURHAM

A HENRY HOLT REFERENCE BOOK
Henry Holt and Company
New York

A Henry Holt Reference Book
Henry Holt and Company, Inc.
Publishers since 1866
115 West 18th Street
New York, New York 10011

Henry Holt ® is a registered
trademark of Henry Holt and Company, Inc.

Copyright © Text 1996 by Roger Osborne and Don Tarling
Design and illustrations, 1996 by Swanston Publishing Limited
All rights reserved.
Published in Canada by Fitzhenry & Whiteside Ltd.,
195 Allstate Parkway, Markham, Ontario L3R 4T8.

Library of Congress Catalog Card Number: 95-79328

ISBN 0-8050-4552-X

Henry Holt books are available for special promotions and
premiums. For details contact: Director, Special Markets.

First Edition—1996

Printed in Great Britain
All first editions are printed on acid-free paper. ∞

10 9 8 7 6 5 4 3 2 1

Preface

We have two aims in writing and compiling *The Historical Atlas of the Earth*. First, to present our interpretation of the current state of knowledge of the Earth's history, and in particular its changing geography. Second, and just as important, we hope to bring the subject of Earth Sciences or geology to a wider audience.

Geologists have, for the last two hundred years, been gathering information about the history of the Earth. The overarching theory of plate tectonics gave a dramatic new impetus to the study of the Earth, comparable to the effect of evolutionary theory on the biological sciences. The resulting mass of information about the Earth's changing history has also, in our view, provided a new and exciting way of presenting the study of the Earth to a general readership. Too often geology is presented as a method for discovering what lies beneath the Earth's surface, through the study of static rock outcrops. But in fact geologists have always been engaged in continual acts of imagination, recreating ancient environments and past worlds through the interpretation of the evidence of the present.

Once it was known that the continents were moving and had moved, it became logical to ask where they had been at different times in the past. This led directly to the invention of a whole new discipline within Earth Science, known as paleogeography – the study of ancient geography. Paleogeographers have been able to painstakingly reconstruct the movements of the continents throughout their history. They have shown how these movements not only account for the geological events of the Earth's history, but how the mechanism that drives them is essential to the formation and maintenance of the atmosphere, the oceans and the development and sustenance of life. *The Historical Atlas of the Earth* enables us to use maps showing the changing geography of the Earth to explain the history of our planet.

During the course of writing this book, we have come to understand the size of the task before us. Compiling the entire geological history of the Earth in one volume has involved a certain amount of selectivity, though we have aimed to be as comprehensive as the available information would allow. Readers will find that some information is repeated in different parts of the book. This is deliberate, since we wish to make each double-page a complete summary of the topic being covered – without the necessity of absorbing all the preceding material.

The paleogeographic maps in this atlas are drawn from the latest research data, re-presented in a form that is attractive and informative for general readers and geologists alike. Paleogeography is an exciting and rapidly developing subject. So while our maps are as up-to-date as possible, they are certain to be superseded in time, although we believe that the general approach we have adopted will continue to excite and interest readers.

We have built on the work of hundreds of scientists and have been helped by numerous friends and colleagues in compiling this atlas. In particular we would like to thank Professor C R Scotese of the University of Texas at Arlington, Professor N N Ambraseys at Imperial College, London and Dr Mike Benton at the University of Bristol for advice and information. Above all we must thank Dr G A L Johnson at the University of Durham. As well as contributing to the book, Tony gave us his enthusiasm and encouragement when the atlas was no more than the glimmer of an idea.

While we fondly hope that some future Earth Scientists will be able to look back on *The Atlas of Earth History* as their inspiring introduction to the subject, our first aim is that readers find it an enjoyable and informative way to learn about the history of their planet.

Roger Osborne, Scarborough
Don Tarling, Plymouth

Contents

Introduction

The essential basis of any history is the ability to place events within a time frame. A history of the Earth demands a longer time-scale, but is nonetheless governed, driven and given meaning by reference to the passage of time. In fact, the proper study of geology could only really begin once the Earth was known to have existed for many millions of years.

The discovery of 'deep time' in the late 18th and early 19th centuries was a revolutionary event in the the study of the Earth. Educated opinion had until then followed a biblical time-scale and reckoned the Earth to be about 6,000 years old. Geologists had already realized that, as many geological processes happened at such a relatively slow rate, the formation of the Earth could not be accommodated within the time available. Once the evidence that the age of the Earth should be measured in millions, not thousands, of years became overwhelming, the way was open for geologists to begin to detail its history.

The impact of deep time on evolutionary theory was even more immediate. Charles Darwin knew that transmutations of species often took place over many generations, demanding millions of years of history for the diversity of life he saw on Earth to have evolved by natural selection. The arguments that geologists like Charles Lyell put forward for an Earth millions of years old were an essential underpinning of Darwin's work, and overcame what was a potentially serious scientific obstacle to the acceptance of evolutionary theory.

While the Earth was formed about 4,600 million years ago, the oldest known rocks found on the Earth are about 3,900 million years old – that is therefore the length of the geological time-scale. Such lengths of time are beyond the understanding of human perception. We simply cannot envisage the passage of one million years, far less thousands of millions. This is unsurprising since human beings themselves have only existed for a tiny proportion of the Earth's history, while recorded human history is a small portion of that. One person's lifespan is immeasurably small on the geological time-scale.

The incomprehensible vastness of this endlessly regressing time-scale, and the apparent slowness of geological processes can make geology seem a remote science, with little relevance to the time-scale of human affairs. Yet the evidence of our daily lives shows the opposite. The Earth has had a long history, but not a quiet one. Even in its most tranquil times it is a turbulent place. For humans geology is an everyday event as well as a historical study. We may not all have witnessed a volcanic eruption, or an earthquake or a landslide, but we have all read about them in our newspapers and seen them on film. Volcanoes and earthquakes happen instantly and with little warning. They can transform the landscapes around them in a matter of hours. More gradual processes like sea level and climate changes have happened with surprising rapidity within the span of human history. Britain was joined to Europe from 10,000 to 6,000 years ago, allowing the migration of peoples. Alaska was joined to Siberia via the Bering land bridge, and it was along this route that the first humans came to North America.

The age of the Earth and the birth of geology

It was Archbishop Ussher, master of Trinity College Cambridge, in 1654, whose studies led him to the conclusion that the world had been created on 23rd October 4004 BC at noon. That date was based on calculations of astronomical cycles and analysis of the numbers of generations in the Old Testament. As long as the simultaneous creation of the Earth and its inhabitants by a divine miracle was held to be inalienable (and it was believed that geological processes happened very rapidly), then such a short time-span for the Earth was plausible. In the course of the 18th century it became apparent that layers of rock, or strata, must have formed very slowly over great lengths of time. While it was thought possible that fossilized animals might represent still-living groups, the process of fossilization was nevertheless understood to be very gradual. The requirements for a greater timespan led naturalists to devise more scientific methods for dating the Earth. Among the first was Buffon, who measured the time taken for iron spheres to cool, and compared this with an object the size of the Earth. He estimated the Earth to be 75,000 years old, beginning a trend which saw the age of Earth steadily increase over the next 150 years.

Some pre-Christian thinking had held the age of the Earth to be infinite – a concept that was anathema to literal believers in the Bible. But the idea was revived by geologists at the end of the 18th century to accommodate their increased understanding of the gradual nature of geological change. These geologists saw that there were great eruptions and earthquakes happening all the time. But they also saw slower processes of deposition and erosion. They understood that, though catastrophic geological events were happening all the time, they occurred on a regular basis, and this regularity would need to recur over a vast length of time to account for the present state of the Earth. In this sense, catastrophic episodes repeated over and over in regular cycles (there are about 150 major earthquakes and 20 volcanic eruptions a year), become normal, or geologically routine.

James Hutton proposed that the Earth went through regular cycles, in much the same way as the planets orbit the Sun. For Hutton the task of geologists was to ascertain the laws that governed the regular behaviour of the Earth, rather than to try to determine its age, or its linear history. In his most famous words: 'If the succession of worlds is established in the system of nature, it is in vain to look for anything higher in the origin of the Earth. The result therefore of our present enquiry is, that we find no vestige of a beginning, – no prospect of an end.' This last sentence, which closed Hutton's great work *The Theory of the Earth,* is taken to indicate the necessity of deep time. But it is more likely that Hutton was expressing his belief in the cyclic nature of time. The previous sentence carries another evocative phrase, 'the succession of worlds', which established the idea on which modern geology is built, that the Earth has gone through massive changes which have altered its surface radically since it was formed. This idea replaced any notion that the Earth had been made fully formed, and that continuing processes of erosion by seas and rivers, and volcanic eruptions, were merely tinkering with a static creation. Hutton believed, as Newton had done, that the world was set in motion by a divine hand, and that motion was guided by natural laws. We are merely living in the latest result of the actions of these laws – in the latest of a succession of worlds.

Although all 19th-century geologists believed that the age of the Earth must be measured in millions of years there was considerable dispute over the actual number – or even the order of magnitude. Charles Lyell, Darwin's mentor, believed that not only did geological processes happen slowly, but they happened at exactly the same speed and in exactly the same way at every point in the past. This doctrine of uniformitarianism gives an avenue for measuring the age of the Earth. If things happened at the same speed throughout the past, then they are still doing so at the present, and the rates are therefore measurable. These so-called hour-glass methods were based on a number of different processes.

In the 18th century William Halley had measured the salinity of large salt water bodies at regular intervals, and calculated the length of the time when there was no salt in them. The method was refined by later scientists, and by 1899 gave the result of 99 million years. Measurement of the rate of accumulation of sediment was compared with the dimensions of the thickest rock sequences. This gave answers varying from 27 to 1,600 million years – again of the right order. But these methods were overtaken by estimates of heat loss by Lord Kelvin. He calculated the age of the Earth to be 20 to 30 million years old, and certainly not more than 100 million years old, since before then the Earth would be too hot to support life. Kelvin's reputation as an experimental scientist established his results as the benchmark.

The subsequent discovery of radioactivity led to the realization that the Earth had its own continuous heat source. This overthrew Kelvin's assumptions of a progressively cooling Earth, and made his calculations irrelevant. In addition radioactive dating provided a better method for measuring the age of rocks, and proved his predecessors to have been more accurate. Radioactive dating works by measuring the proportions of certain chemical isotopes in rock formations. Because these isotopes decay at a known rate, the age of the rock can be calculated. The advantage over other methods is that rates of radioactive decay are not much affectd by changes in temperature and pressure, though anomalies can occur through isotopes 'infecting' the rock from outside. Radioactive dating has shown the oldest rocks on Earth to be about 3,900 million years old.

Fossils and the geological timescale

The most remarkable achievement of 19th-century geology was to build a system for dividing up the geological timescale and classifying rocks within it, without knowing the actual ages of any of the rocks.

The fact that this system survived the introduction of radioactive dating largely intact is a glowing testament to the collective work of those scientists. The geological timescale, with its divisions into eras, and periods and systems, is based purely on the relative, rather than the actual, ages of rocks. This immense three-dimensional jigsaw puzzle was put together by study of the fossilized remains of plants and animals, particularly marine animals, and of the spatial relations that rock formations bear to one another.

The initial supposition that rocks should lie neatly on top of one another in the order in which they were formed – the oldest at the bottom, the youngest at the top – was rudely overturned by the evidence. Any of thousands of cliff faces reveals folded, faulted and contorted rock strata. When rock beds are carefully followed they are often found to turn upside down, or to lie on top of rocks that they are underneath in other locations. Once geologists began the massive task of unravelling the geometric relations between rocks, they realized that the fossils preserved in sedimentary rocks would be an essential tool in understanding the history of the individual rock formations, and of the Earth's crust in general.

Fortunately for geologists some groups and species of marine invertebrates change the shapes and structures of their shells over fairly short timescales. Some – particularly those which are free-floating rather than static – also spread rapidly over large areas. When these two characterisitics are combined, fossils of that animal are an invaluable tool in deciphering relations between rock formations. Changes in these *zone fossils* or *marker fossils* can pinpoint the relative ages of rock strata with great accuracy, even where the rock formations are separated by great distances.

The divisions of the geological time-scale are based on the changes in fossils, usually marine invertebrates, found in rock strata. This is why the great extinctions occur at the end of geological periods or eras. The four great eras are the Precambrian, the Paleozoic ('ancient life'), the Mesozoic ('middle life'), and Cenozoic ('new life'). The boundary between each of these is marked by a drastic change in life forms – or, more accurately, a change in life forms preserved as fossils. The boundary between the Mesozoic and the Cenozoic, 65 million years ago, coincides with the extinction of the dinosaurs and ammonites, respectively the dominant land animal and the most important fossil group of the Mesozoic era. This allowed the subsequent domination of the land by mammals and birds. The Paleozoic came to an end 250 million years ago with the extinction of a number of groups of marine invertebrates, including the trilobites, the characterisitc fossil of Paleozoic rocks. The boundary between the Precambrian and the Paleozoic was thought by early geologists to mark the beginnings of life; in fact it marked the start of marine animals developing shells and external skeletons. The soft-bodied animals that came before left virtually no fossil traces of their existence so the Precambrian was simply defined as its name suggests – all that came before the Cambrian period. Unfortunately the Precambrian – running the age of the oldest rocks, to 550 million years ago – covered 85 per cent of the geological time-scale. It is only through modern techniques of electron microscopy and radioactive dating that geologists have been able to use those traces of life found in the Precambrian to work out some sort of chronological system for the huge timespan of the era.

Within each era the geological timescale is divided into periods. Again, these are based on changes in fossils, but also on sudden breaks in rock types. The periods are generally named after areas where early geologists were working, and where rocks of that age were commonly found. Most of this early work was done in northern Europe and North America, so the boundaries between the periods are marked by sudden changes in fossils or rock types in these regions. Elsewhere in the world, changes may not occur suddenly at those times, and the boundaries can only be defined by radioactive dating. In general though the system holds well across the world, and throughout the geological record.

Plate tectonics and continental drift

While the discovery of deep time enabled the establishment of geology as a subject, a second revolution in Earth sciences happened with the acceptance that the continents had moved around on the Earth's surface. In retrospect, we could say that the evidence for continental drift gradually became overwhelming, until all objections to the theory were pushed aside, and all geologists fell into line. In practice, a few scientists in the early part of this century, principally Alfred Wegener, a German meteorologist, pushed forward the idea in the teeth of important objections based on current scientific evidence. Even those indicators which seem to us now to confirm the idea of continental drift were quite plausibly explained in other ways by objectors. For example, evidence of past glaciation in tropical regions could be explained by ancient ice ages which gripped the whole Earth.

Even for geologists used to dealing with the massive forces of the Earth shown in volcanic eruptions, earthquakes and the uplift of mountain chains, the movement of whole continents was a phenomenon of a different physical order. In fact the hesitation of geologists in accepting continental drift was based not so much on the lack of evidence, but on the lack of a plausible mechanism for pushing enormous landmasses around on the Earth's surface. The mechanism which is now generally accepted is known as *plate tectonics*. Our present understanding is that the layer of the Earth below the crust – the mantle – is partially molten. Heat sources within the mantle (radioactive elements) set up huge convection currents, which in turn induce flows across the upper part of the mantle. Since the solid crust sits on the mantle it is affected by these flows. The solid crust is in fact made up of a series of rigid plates (six major and many more minor) which move in response to the mantle currents beneath – hence continental drift. Where plates are moving towards each other, one will be pushed under the edge of the other and back down into the mantle. These destructive plate boundaries are where most of the Earth's geological activity takes place.

The plates which make up the ocean floor are contstantly recycled in this way, while the continents are permanent, and have been gradually growing through the addition of more crust at their edges over the 4,000 million years of the Earth's geological history. Plate tectonics remains a theory, since the Earth's mantle is fairly inaccessible, and the movements within it are very gradual. But the theory has been remarkably successful in accounting for the geological phenomena of the past and the present. Work on the origins of the atmosphere and oceans, and comparisons with the geologically 'dead' Moon, have also shown that the mechanisms outlined in the theory are vital for the continuation of the life of, and the life on, the planet.

The Earth's changing geography

Continental drift reveals that Hutton's phrase 'a succession of worlds' has more meaning than he could have known. The familiar map of the world, that underlies every thought we have about the geography of the Earth, is shown to be merely a snapshot in time. The continents have never been arranged in their present form before, and they never will be again. The following pages tell the story of the Earth's history through its changing geography. But the movements of the continents are not simply a matter of changing spatial arrangements. The creation of continents, the separation and collision of landmasses, the movement through different latitudes, the diversion of ocean currents, the creation of new oceans – all these have had a massive effect on the geology, climate and lifeforms of the Earth throughout its history. It is no exaggeration to say that the changing geography of the Earth is its history, since it has been shown to be the starting point for all the events of the Earth's past.

The reconstruction of the Earth's geography is known as *paleogeography* ('ancient geography'). Evidence used by paleogeographers is taken from fossils, rock types, internal and external rock structures, and especially from paleomagnetism ('ancient magnetism'). The realization that some rocks carry the record of the Earth's magnetic field at the time they were formed, was a breakthrough in confirming the theory of continental drift. Since the latitudes of ancient continents at different times can be calculated from the magnetism in their constituent rocks, paleomagnetism is the most important tool in mapping the positions and movements of continents. Paleogeography is a young science, having only started in earnest once continental drift was generally accepted in the 1960s. In the 30 or so years since, remarkable progress has been made in mapping the Earth throughout its history. In addition, the geography of the Earth has come to be seen as the key to understanding its past, and possibly its future, development.

Part One
The Cosmos

Big Bang to Stellar Space

Meteorites, Asteroids and Comets

Formation of the Planets

The Moon

Motions in the Earth

The Earth's Magnetic Field

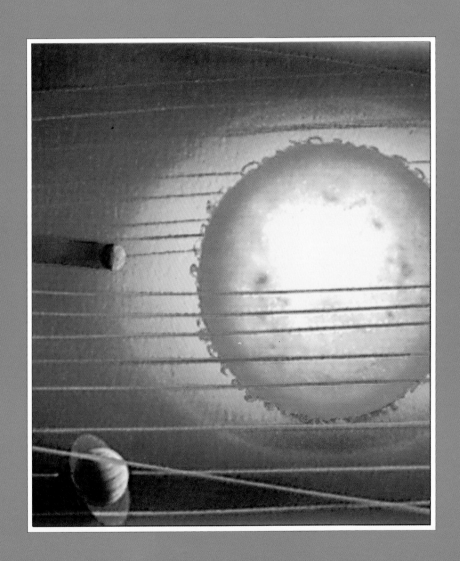

Big Bang to Stellar Space
The birth of the universe

Cosmologists believe that the universe began as a single point, infinitely hot and infinitely dense. This point exploded outwards in what we call the Big Bang, which was the start of the universe. The question of what existed before Big Bang is therefore difficult to answer. One interpretation of relativity theory indicates that time itself would not have existed before the Big Bang. Another theory is that the universe oscillates over billions of years: it expands from a Big Bang, as it is now doing, but then at some point it begins to contract. Eventually it implodes into an infinitely dense point, the Big Crunch, another Big Bang occurs and the cycle begins again.

If there was a Big Bang, where did the material in the Big Bang come from? There is now a theory, known as the Free Lunch theory, which proposes that under the right conditions you can get something from nothing – in other words, the matter of the universe could spontaneously have come into existence. This is by no means generally agreed.

1 The moment of the Big Bang, 15 billion years ago. The universe is concentrated into a single point of infinite density and heat.

2 After a few billionths of a second, the universe has expanded from the size of an atomic nucleus to the size of a basketball.

3 The universe is one millionth of a second old. It is an exploding ball of fire, which has expanded to 100 billion kilometres in radius. The ball contains subatomic particles – electrons, protons and neutrons – and neutrinos, but it is too hot for formation of atoms.

4 One minute after Big Bang. The universe is now a million billion kilometres across. The hydrogen nuclei (containing one proton) formed in the Big Bang are being converted to helium nuclei in a gigantic thermonuclear reaction. The temperature is several billion degrees – too hot for atoms to form.

5 Half a million years after Big Bang. The temperature is around 4,000°C, cool enough for atoms to have formed from the subatomic particles. The overall temperature is that of the surface of the Sun at present. As atoms form, matter begins to come together under gravitational attraction. The intense radiation of the Big Bang explosion fades and the universe darkens.

6 Several billion years on. Matter, which is in the form of huge gas clouds, begins to contract. Stars are formed and galaxies are born.

7 The present, 15 billion years after the Big Bang. The galaxies of stars in the universe are grouped together in separate superclusters, which are 100 to 400 million light years apart, with vast areas of darkness between them. There is some residual background radiation from the Big Bang. The universe continues to expand. It may be that the dark matter that makes up 99 per cent of the universe will eventually halt and reverse this expansion through its own gravitational attraction.

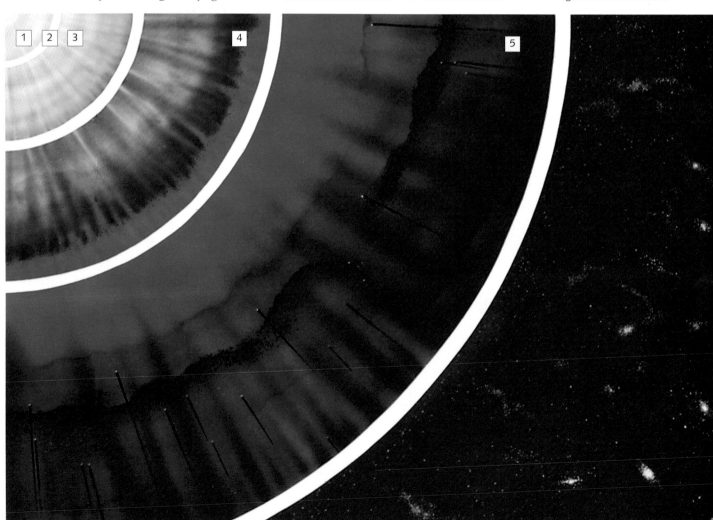

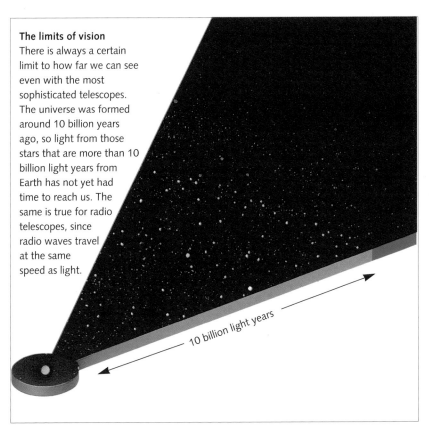

The limits of vision

There is always a certain limit to how far we can see even with the most sophisticated telescopes. The universe was formed around 10 billion years ago, so light from those stars that are more than 10 billion light years from Earth has not yet had time to reach us. The same is true for radio telescopes, since radio waves travel at the same speed as light.

10 billion light years

Life cycle of a star *(below)*

The formation of a star begins with a cloud of nebular material. The cloud begins to collapse inward under the force of its own gravitational pull and the temperature begins to rise. Stars form as the temperature reaches levels at which nuclear fusion can take place. Only some of the many bodies are big enough to have the internal gravitational pressure to trigger such reactions. The stars shine as energy is released in the form of heat and light. The remaining gas in the nebula is then blown away and a star cluster remains.

The further evolution of a star depends upon its mass. A star like our Sun enters the stable part of its existence and continues to radiate for thousands of millions of years. When its hydrogen begins to run down, the star then expands and becomes a Red Giant. The layers of gas surrounding the star are dissipated and the remaining core continues to shine feebly as a White Dwarf, before finally losing all of its heat to become a cold, dead Black Dwarf.

If the star has greater mass it evolves much more quickly. After a stable period it becomes a Red Supergiant and may explode as a Supernova. It may end its life as a neutron star or a pulsar. If its mass is very large it could implode and end up as a Black Hole.

6

7

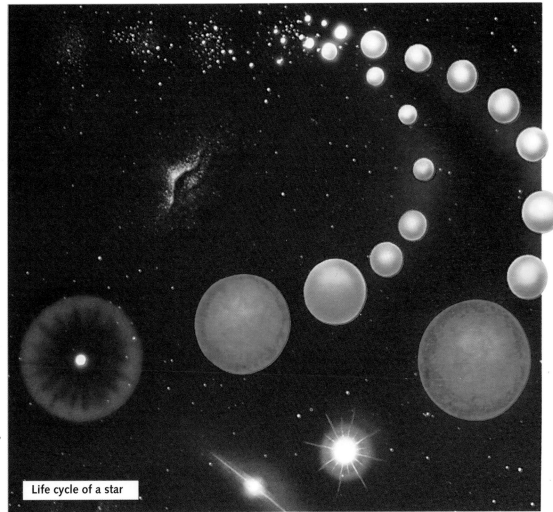

Life cycle of a star

Meteorites, Asteroids and Comets
Clues to the origins of the Earth

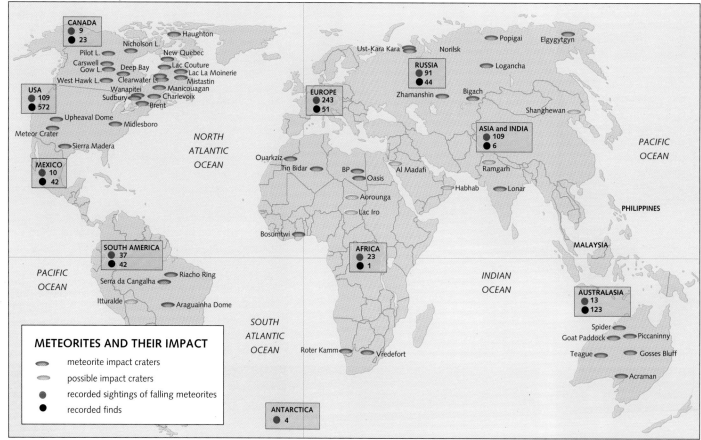

METEORITES AND THEIR IMPACT

- meteorite impact craters
- possible impact craters
- recorded sightings of falling meteorites
- recorded finds

Geologists look upward as well as downward for clues about the history and composition of the Earth. Before the advent of space flight, meteorites were the only extra-terrestrial material available to us. Though most burn up in the atmosphere, between 30 and 40 land on the Earth and are found and recovered each year. Meteorites provide important clues to understanding the formation of the Earth and the other planets. They are probably of similar composition to the fragments that originally combined to form the inner planets. About a quarter of meteorites found on Earth are metallic. They contain mainly iron and nickel and may be made up of the cores of fragmented asteroids. The remainder are stony meteorites and are probably from the mantle or crust of asteroids.

Meteorites come from the asteroid belt, or from the far reaches of the solar system along with those other extra-terrestrial visitors, comets. As the new planets accreted from the solar nebula, they attracted debris from their local area, by gravitational force. Most of this was accreted into the planets, or ended up orbiting them, like the material in Saturn's rings. A certain amount of debris was, however, left over as meteorites, asteroids and comets.

Meteorites and their impact

Millions of meteors enter the Earth's atmosphere every day. Most of these are very small and are burned up by the atmosphere – this burning is sometimes seen as the trail left by a 'shooting star'. Their remains reach the ground as space dust – several thousand tonnes settle on the Earth's surface every year.

A meteor becomes a meteorite when it hits the Earth's surface – 500 or so do every year (*above*). Although meteors weighing less than a tonne tend to break up into smaller fragments, larger meteors, that can withstand the journey through the Earth's atmosphere, can strike the Earth at close to their original entry speed of 250,000 km per hour (155,000 mph).

Meteor crater, Arizona
This crater, near the town of Flagstaff in the Arizona desert *(below)*, is 1,300 metres (4,200 ft) across and over 180 metres (600 ft) deep, with a raised rim 40 metres (130 ft) high. Rocks in and around the crater show signs of intense heat. The impact that caused this crater occurred about 10,000 years ago.

Meteor Shower

Spectacular meteor showers *(left)* are a regular feature of the night sky. They occur when the Earth passes through the orbit of a comet. Particles of dust that have been left in the comet's wake are burned up in the Earth's atmosphere, giving the effect of hundreds of 'shooting stars'. The particles are much too small to become meteorites. Increased knowledge of comet paths has enabled astronomers to predict meteor showers.

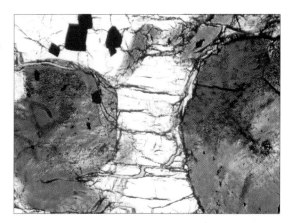

Comets

Comets originate in the far reaches of the solar system, in a region known as the Oort cloud where ice and dust have accumulated. Very few comets approach the inner part of the solar system, where they would be visible. Comets are typically balls of ice with rock particles up to 10 km (6 miles) in diameter. Those comets that we can see have a very elliptical orbit that takes them near to and away from the Sun. If one of them approaches the Sun a gaseous tail is formed by the effect of the Sun's radiation. This characteristic tail is what we see, rather than the comet itself.

Meteorite composition

A thin section of a meteorite is shown under a plain light *(top)* and polarized light *(above)*. This technique is used by geologists to analyze the mineral content and structure of rocks. The composition and internal structures of meteorites gives clues to the Earth's composition.

THE ORBIT OF HALLEY'S COMET

Pluto

Neptune

Uranus

Saturn Earth

Sun

Halley's Comet

Halley's Comet

The most famous comet of all is Halley's comet *(right)* – seen here in its 1910 appearance. Its orbit brings it past the Earth every 76 years (its orbit is variable, because the low mass of comets makes them susceptible to gravitational pulls from other bodies). It last visited in 1986, though it was not visible from most of the Earth.

As well as giving us an insight into the workings of the solar system, comets and meteorites play another important part in geologists' thinking. Mass extinctions of life-forms, particularly the disappearance of the dinosaurs, have been one of the great puzzles of our planet's history. There is much speculation that comets or asteroids may have been involved in these events.

Asteroids

The asteroid belt contains a million objects varying from 1 to 950 km (.6 to 580 miles) in diameter which circle the Sun in a belt between Mars and Jupiter – meteorites are the only direct samples we have of material from the asteroid belt. Remote sensing suggests that, while the asteroids nearer the Sun are more metallic, they become progressively more silicon-rich, and then more carbon-rich the further out they are in the belt. This echoes the composition of the planets.

Formation of the Planets
Solid worlds from solar nebula

Ever since Copernicus showed that the Earth and planets revolve around the Sun, there have been widely differing theories of how the planetary system evolved. In recent years planetary probes have brought back data on the atmospheres and surfaces of the Moon, Mars, Venus, Mercury, Jupiter and Saturn. A new infra-red telescope (IRAS) has given us firm evidence of the formation of solid matter around other stars. There may be planetary formations around the nearby stars, Vega and Fomalhout.

Most theories of the birth of the planets start from the idea of an original rotating cloud of dust and gas, held in tension by the opposing forces of rotation and gravitational attraction. At some point gravitational forces became stronger and the cloud began to contract. The cloud flattened into a disc and matter began to be pulled toward the centre. As material gathered more and more in the centre, what was to become the Sun collapsed under the force of its own gravitation. The Sun's temperature rose to 1,000,000°C. It was then hot enough for nuclear fusion, in which hydrogen nuclei combine to form helium nuclei while releasing an enormous amount of energy, to begin.

Pluto *(from Hubble telescope)*
Diameter:14,000 km (8,515 miles)
Mass: 0.8
Gravity: Unknown
Satellites: 0
Day length: 6.49 days
Year length: 248 years
Distance from Sun: 5,886 million km (3,580 million miles)

Earth
Diameter: 12,756 km (7,759 miles)
Mass: 1.00
Gravity: 1.00
Satellites: 1
Day length: 1 day
Year length: 1 year
Distance from Sun: 149 million km (93 million miles)

Mercury *(from Mariner 10)*
Diameter: 4,835 km (2,940 miles)
Mass (relative to Earth): 0.055

The Formation of the Solar System

It is widely belived that the planets were gradually formed as the nebula of gas and dust cooled. Mercury, the planet closest to the Sun, is formed from materials with high melting points, like minerals and metals, which began to solidify at about 1,000°C. As the nebula cooled to 200–700°C, Earth, Venus and Mars were formed by grains of silicon and magnesium minerals sticking together to

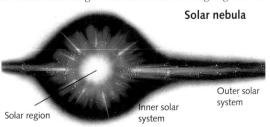

Solar nebula

Solar region

Inner solar system

Outer solar system

form chunks called planetesimals, which pulled in matter by gravitation. Saturn and Jupiter, the outer planets, were formed at much lower temperatures, and consist now of swirling masses of volatile gases and liquids. Their composition closely reflects that of the original nebula.

Saturn *(from Voyager 2)*
Diameter: 120,800 km (73,479 miles)
Mass: 95.1,Gravity:1.7
Satellites: 15
Day length: 10.23 hours
Year length: 29.5 years
Distance from Sun: 1,421 million km (864 million miles)

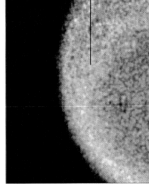

Uranus *(from Voyager 2)*
Diameter: 47,100 km (28,650 miles)
Mass: 14.5
Gravity: 1.03

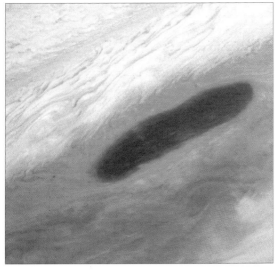

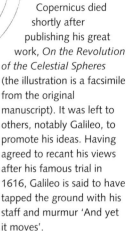

Copernicus died shortly after publishing his great work, *On the Revolution of the Celestial Spheres* (the illustration is a facsimile from the original manuscript). It was left to others, notably Galileo, to promote his ideas. Having agreed to recant his views after his famous trial in 1616, Galileo is said to have tapped the ground with his staff and murmur 'And yet it moves'.

Copernicus manuscript
The theory that the Earth moves around the Sun, not vice-versa, was put forward by Nicolaus Copernicus, a Polish astronomer, in 1542. The theory had already been proposed by Pythagoras among others, but detailed observation of the planets gave it irresistible authority.

Gravity (relative to Earth): 0.38
Satellites: None
Day length: 59 days
Year length: 0.24 years
Distance from Sun: 57.5 million km (35 million miles)

Jupiter *(part of atmosphere, from Voyager 1)*
Diameter: 141,600 km (86,131 miles)
Mass: 318
Gravity: 2.64

Satellites: 15
Day length: 9.83 hours
Year length: 11.99 years
Distance from Sun: 775 million km (471 million miles)

Venus *(computer image)*
Diameter: 12,914 km (7,854 miles)
Mass: 0.815
Gravity: 0.89

Satellites: None
Day length: 243 days
Year length: 0.616
Distance from Sun: 107 million km (65 million miles)

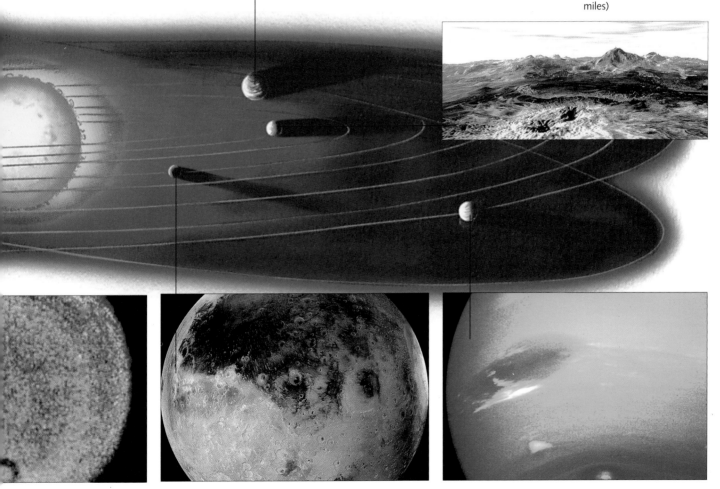

Satellites: 5
Day length: 23 hours
Year length: 84 years.
Distance from Sun: 2,861 million km (1,740 million miles)

Mars *(mosaic image)*
Diameter: 6,760 km (4,112 miles)
Mass: 0.108
Gravity: 0.38

Satellites: 2
Day length: 1.03 days
Year length: 1.88 years
Distance from Sun: 226 million km, (137 million miles)

Neptune *(from Voyager 2)*
Diameter: 44,600 km (27,128 miles)
Mass: 17.0, Gravity: 1.5

Satellites: 2
Day length: 22 hours
Year length: 165 years
Distance from Sun: 4,497 million km (2,794 million miles)

The Moon
Earth's dead satellite

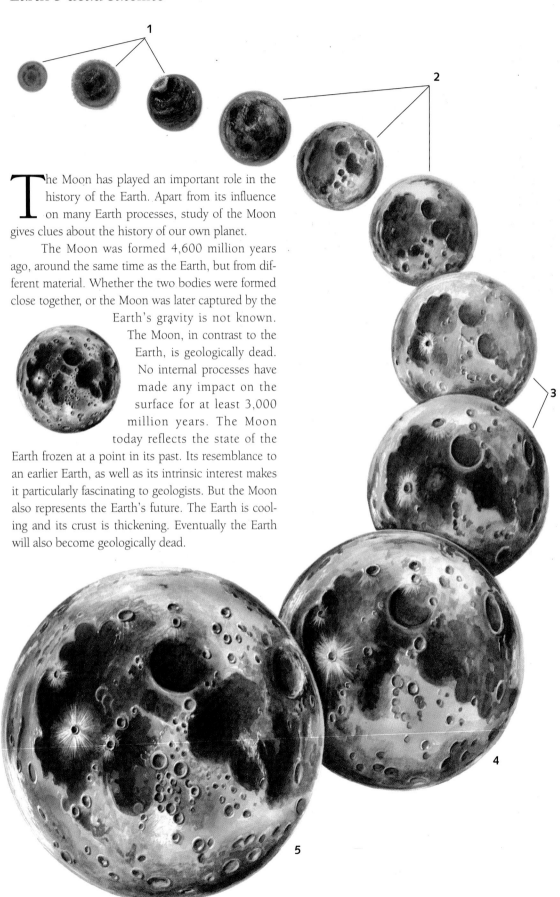

The Moon has played an important role in the history of the Earth. Apart from its influence on many Earth processes, study of the Moon gives clues about the history of our own planet.

The Moon was formed 4,600 million years ago, around the same time as the Earth, but from different material. Whether the two bodies were formed close together, or the Moon was later captured by the Earth's gravity is not known. The Moon, in contrast to the Earth, is geologically dead. No internal processes have made any impact on the surface for at least 3,000 million years. The Moon today reflects the state of the Earth frozen at a point in its past. Its resemblance to an earlier Earth, as well as its intrinsic interest makes it particularly fascinating to geologists. But the Moon also represents the Earth's future. The Earth is cooling and its crust is thickening. Eventually the Earth will also become geologically dead.

The formation of the Moon

1. 4,600 million years ago
After its formation the Moon heated up in much the same way as the Earth. A very thick molten mantle was formed early on, and from this lighter materials floated out to form a crust. A small iron core produced a magnetic field.

2. 4,000 million years ago
The Moon was subjected to intense bombardment by meteorites which created vast craters covering most of the surface.

3. 4,000-3,000 million years ago
Floods of volcanic material from the Moon's interior formed the great *mare,* or seas. These plains of basaltic lava flooded into massive meteorite craters and then solidified. The crust formed into a single plate, much thicker, 120 km (75 miles) and stronger than the multi-plated, mobile, thin crust of the Earth.

4. 3,000 million years ago
Because of its smaller size the interior of the Moon cooled much more rapidly than the Earth. Convection currents in the interior stopped, ending all volcanic activity and terminating the magnetic field. The Moon became geologically 'dead'. There are no volcanic rocks on the Moon younger than 3,000 million years old.

5. 3,000 million years ago to the present
Since the solidifying of the core, activity on the Moon has been confined to the altering of the surface by the impact of meteorites and wind erosion, and moonquakes. The large Copernicus and Tycho craters were formed 800 to 400 million years ago.

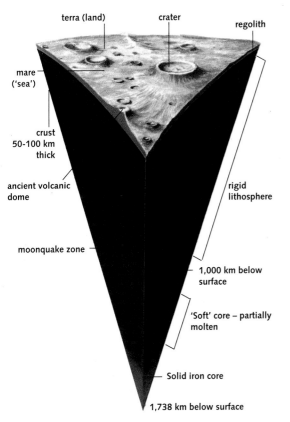

terra (land)　　crater　　regolith

mare ('sea')

crust 50-100 km thick

ancient volcanic dome

moonquake zone

rigid lithosphere

1,000 km below surface

'Soft' core – partially molten

Solid iron core

1,738 km below surface

THE MOON'S GEOLOGY

impact craters >50km

mare plateaus

younger mare

older mare

dark mantles

light terra plains

hilly and furrowed terra

hilly and pitted terra

basin deposits and structures

cratered light terra plains

terra, undivided

terra, densely cratered

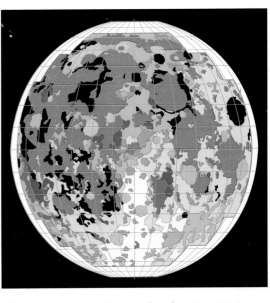

The Moon's geology (*above*) **and surface** (*photo below*)
Because of its lack of atmosphere and water the Moon has no erosion and no sedimentary rocks. The surface consists entirely of volcanic rocks.

The interior of the Moon

The Moon's crust is known as the *regolith*. Ancient parts of the crust are present as mountains and highlands. These are known as *terra*, because early astronomers saw them as pieces of land rising above the flat areas which they called *mare*, or seas. The *mare* are flat plains of lava which flowed into the massive craters of early meteorites. With no water and little wind, all the surface features are caused by volcanic activity or meteorites.

There is now evidence of moonquakes, though their intensity is one billion times smaller than earthquakes. Moonquakes originate at the boundary between the soft, partially molten outer core and the rigid lithosphere 1,000 kilometres (620 miles) below the surface. Quakes tend to increase when the Moon is near to the Earth.

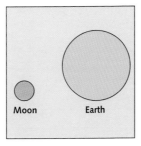

Moon　　Earth

Comparative size

Our Moon is exceptionally large in relation to Earth. But it is not large enough to sustain the heat necesary for geological activity.

Apollo

As a result of the Apollo missions lunar exploration provided detailed photography of the surface, information from instruments placed on the surface, and most important of all, Moon rocks. The photograph shows a Lunar Roving Vehicle and astronauts.

Motions in the Earth
The active forces beneath the crust

The complex history of our planet has only become apparent to us over the last 30 years. In that time the theory of plate tectonics has been accepted, refined and widely applied as a remarkably successful framework for understanding the history of the Earth.

The idea that continents move around on the Earth's surface has existed since the 17th century.

Serious evidence in support of the claims of Continental Drift was put forward by a German meteorologist Alfred Wegener in 1912. Initially greeted with scepticism, the idea of continental movement is now accepted. The science of plate tectonics shows *how* the continents move, and explains that our Earth is a dynamic entity with a history of turbulence, destruction, creation and upheaval.

Convection cells (*below*)
The plates of the Earth's crust are pushed around on the surface of the Earth by the movement of convection currents throughout the mantle, subjecting every part of the crust to moving forces from beneath.

The crust
The Earth's crust is made up of a series of rigid plates, which move around in response to forces in the mantle. The crust-mantle boundary is defined by a change in the behaviour of seismic waves as they move into the more plastic mantle. Continental crust is 34 km (21 miles) deep on average while oceanic crust is 5 km (3 miles) thick.

Mantle material cools and moves sideways in 'convection cell'

Continent crust

Crust

Mantle

Core

Outer core

Oceanic plate

Crustal plate 'floats' sideways on mantle curent

Two plates move towards each other. One is subducted back into the mantle on falling convection current

Material cools as it reaches outer mantle

The mantle
The upper 700 km (435 miles) of the mantle is fairly rigid; the lower 2,200 km (1,367 miles) flows plastically. Convection cells form in the lower mantle and affect the upper portion and the crust.

Mantle

Hot material rising through mantle

Cold dense material falls back through mantle

Mantle material heated here

2,900 km (1,800 miles)

The outer core
Currents in the liquid outer core drive the Earth's magnetic field. Its composition is iron with 10 per cent nickel.

Hot outer core

The heat inside the Earth
We have long known that below the crust the Earth is hot – the evidence is there in volcanoes and hot geysers. But why is the centre of the Earth hot, and where does the heat come from? The answer lies in understanding what happened when the Earth was formed. The Earth has existed for more than 4,600 million years. It was probably formed by small pieces of material, known as planetesimals, accumulating or accreting in space. These initially came together to form a homogenous, undifferentiated mass. As the cold planetesimals accumulated, the planet began to heat up and then separate into distinct layers.

This heating of the new planet was caused by two factors. First, the impact of new bodies striking the planet gave out heat. Second, the process of the Earth contracting, and literally squeezing itself into a smaller space, itself caused heating. The increase in temperature was about 1,000°C. The Earth also has a constant source of heat through the disintegration of radioactive elements like uranium and thorium. Although these are comparatively scarce, the heat they generate is locked into the Earth and accumulates very high temperatures.

The inner core
The inner core is solid iron, probably crystallizing out from the liquid outer core. Our knowledge of the Earth's interior comes from seismic studies.

5,100 km (3,170 miles)

Inner core

North America: a turbulent history

The continent of North America has moved tens of thousands of kilometres over the Earth's surface during the last 500 million years, giving a striking example of continental drift. In the Precambrian period it was located close to the South Pole at nearly 90° to its present alignment (1). It shifted north-west to straddle the equator during the Paleozoic era, moving slowly eastwards (2–5). Since the Carboniferous era it has rotated anti-clockwise and has moved north-west (6, 7), reaching its present position (8). It is now moving westwards and still rotating anti-clockwise. This map shows North America in isolation, but in fact it has ripped away from and crashed into other continents. It has grown by accreting pieces of crust from other continents and from the volcanic eruptions which result from its movements.

Crustal plates, volcanoes and earthquakes

The boundaries between the plates which make up the Earth's crust are the areas where most geological activity takes place. Earthquakes and volcanic eruptions are centred on plate boundaries where plates are moving towards each other. These are known as destructive boundaries.

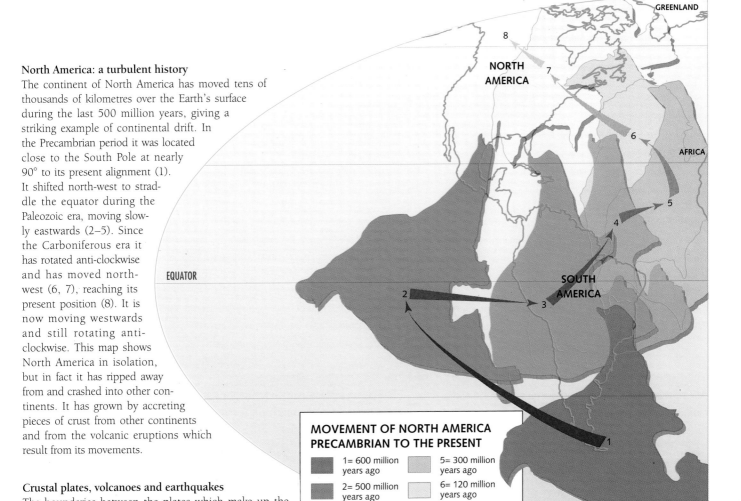

MOVEMENT OF NORTH AMERICA PRECAMBRIAN TO THE PRESENT

1 = 600 million years ago
2 = 500 million years ago
3 = 450 million years ago
4 = 400 million years ago
5 = 300 million years ago
6 = 120 million years ago
7 = 50 million years ago
present position and modern coastline

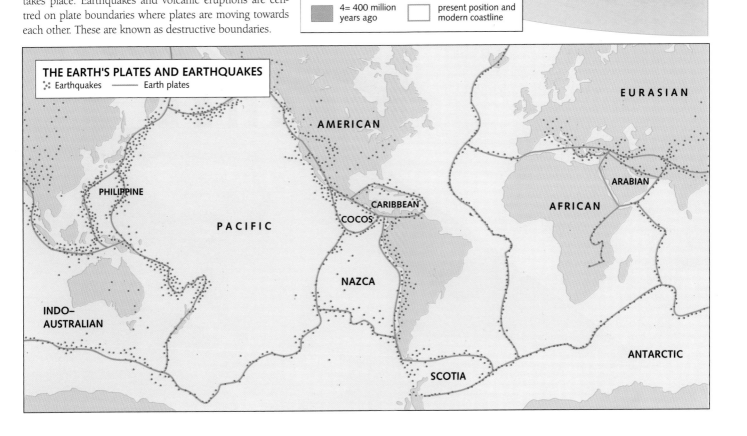

THE EARTH'S PLATES AND EARTHQUAKES
Earthquakes — Earth plates

The Earth's Magnetic Field
The dynamo in the core

When it is first formed every rock is slightly magnetized because of the iron that all rocks contain. The direction of this magnetization is 'locked' into the rock, and is a direct record of the direction of the Earth's magnetic field at the time when the rock was formed.

It was recently discovered that, throughout its history, the Earth's magnetic field reverses at irregular intervals – the South Magnetic Pole becomes the North Magnetic Pole, and vice versa. These inversions are recorded in the Earth's rocks. In addition, the vertical angle of the magnetic field in a rock can tell us what latitude the rock was in when it was formed, information which helps in plotting the positions of the continents throughout their history. The study of the magnetization of rocks is now a vital tool in the reconstruction of the past.

The Earth's magnetic field

All magnetic materials lose their magnetism above a certain temperature, known as the Curie Point. For most rocks the Curie point is about 500°C. As the rocks of the Earth's interior are hotter than this, the Earth cannot be permanently magnetized, like some giant bar magnet hanging in space. Its magnetic field must therefore be continuously generated in some way. The most plausible mechanism for this is the presence of the convection currents in the molten iron of the Earth's core, which form a kind of self perpetuating dynamo.

A feature of the Earth's magnetic field is that the direction of polarity of the magnetization can switch fairly rapidly, over a period of a few thousand years. The Earth's magnetic poles are so close to its rotation poles, because the rotation of the Earth influences the direction of the currents of molten iron in the core, which in turn dictates the direction of the magnetic field. The Earth's magnetic field is currently 12° off from the rotational poles. The exact position of the magnetic pole changes constantly, so navigation charts using magnetic data have to be updated annually.

Magnetization of rocks

All materials are magnetized when they cool to below their Curie point. They then carry the magnetic orientation of the Earth's field at that time. Igneous rocks, formed from cooling magma, therefore show the direction of the Earth's field at the point when they solidify into rock. In sedimentary rocks the process is more subtle. Provided there are tranquil conditions, magnetized particles of sediment will orientate themselves in the direction of the prevailing magnetic field. As they fall slowly through water the tiny particles of magnetized mud and sand will line up. When the sediment forms into rock this direction is preserved. Unfortunately rocks are often reheated by the turbulent events around them, and their magnetization can be destroyed and reformed many times. Geologists have to ensure that the magnetization they are measuring is of the original field, when the rock was first formed.

The history of some rock formations is so complex, with repeated metamorphic episodes, that their original direction of magnetism will never be known.

Sea-floor spreading

Maps of sea-floor rock ages show that rocks increase in age in parallel bands, with distance from central ocean ridges. The patterns are remarkably regular, showing a certain stability in the currents in the mantle. This sea-floor spreading is the way in which the world's ocean basins are formed.

Magnetic inversions in the North Atlantic

As new rocks are formed at the mid-Atlantic ridge they are magnetized in the direction of the Earth's magnetic field at the time. The new rock is then continually pushed away from the central ridge. As a magnetic inversion occurs the next band of newly formed rock is magnetized in the opposite direction, but the rocks that are already formed do not switch their direction of field. Magnetic field inversions occur roughly every half million years, and take a few thousand years to complete. The bands of magnetized rock therefore indicate the age of the rock. The symmetry of the patterns on either side of the ocean ridge shows that North America and Europe have each moved away from the central ridge at the same rate over the past 150 million years.

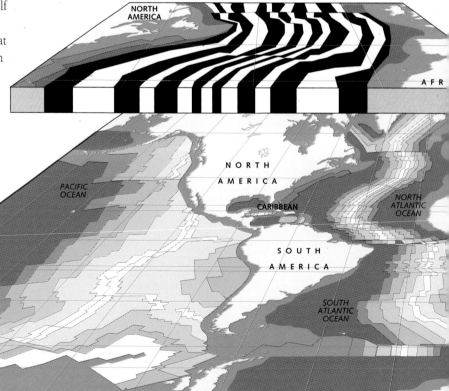

Magnetic readings

The needle of a compass points in the direction of the Earth's magnetic field. At the poles the needle points directly up (or down). At the Equator the needle is horizontal. At other latitudes the needle points at an angle to the Earth's surface. This angle can be used to discover the latitude of a continent when a particular rock was formed – the direction of the Earth's magnetic field is locked into a rock when it is formed.

Sedimentary rocks

Most sediments are formed by particles of rock falling through water *(below)*. During this process, tiny particles of magnetized iron in the grains of sediment become aligned in the direction of the Earth's magnetic field. This alignment is preserved as the sediment solidifies into the rock.

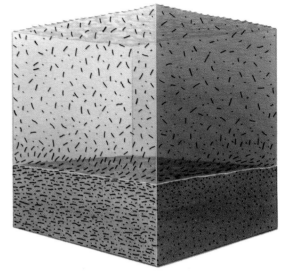

The Mid-Atlantic Ridge

This colour-enhanced image of the Mid-Atlantic Ridge shows variations in gravity measurements. This shows the different thickness of the oceanic crust at the central ridge. The ridge forms as lava spills onto the surface and cools into solid rock – in the same way as land volcanoes are built up from cooled lava.

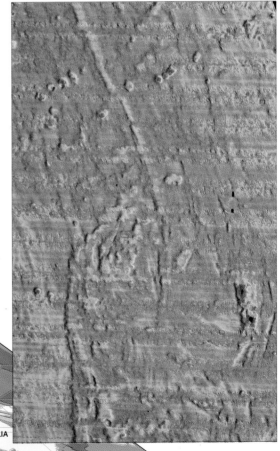

THE AGE OF THE SEA-FLOOR

- 0–5 million years old *(Pleistocene-Pliocene)*
- 5–21 million years old *(Miocene)*
- 21–38 million years old *(Oligocene)*
- 38–52 million years old *(Eocene)*
- 52–65 million years old *(Paleocene)*
- 65–140 million years old *(Cretaceous)*
- 140–160 million years old *(Late Jurassic)*
- 160–190 million years old *(Early Jurassic)*

N

W

E

S

EURASIA

PACIFIC OCEAN

AFRICA

ARABIA

INDIAN OCEAN

AUSTRALIA

ANTARCTIC

Part Two
Making and Dating the Earth

The Making of the Continents
The beginnings of a solid crust

Picture the Earth 4 billion years ago as it might appear from an orbiting satellite. Countless volcanoes are belching forth steam and gases. Spectacular collisions of giant meteorites with the surface of the new planet are common. There is no clear differentiation between continents, oceans, mountains – only small areas of land covered in rocky plains and sand dunes. There is of course no sign of life – no greenery, not even algae or single-celled amoebae. The atmosphere has little or no oxygen and is composed mainly of ammonia and methane. Ultra-violet light from the Sun comes through the atmosphere scorching the surface of the planet. From this unpromising start the planet Earth developed sophisticated chemical and biological systems that have allowed a huge variety of life forms to thrive. As will become clear, the geology and biology and chemistry of the planet are intricately linked, and it is the geology that came first.

The Earth's Crust

For the first billion years of its existence the Earth left no decipherable geological record. It is likely that pieces of crust formed through cooling of the mantle, but that conditions were so unstable that they were continually absorbed back into the molten rock underneath. This cycle of formation and melting went on until about 3,800 million years ago. At that time the surface of the Earth was cool enough at its surface for pieces of crust to be able to resist reabsorption into the molten mantle.

The heat in the interior of the Earth has kept the outer core and most of the mantle molten. These floated upwards and formed the crust. The crust is a solidified skin floating on the molten mantle.

Continental Crust

Two quite distinct types of crust on the Earth's surface – continental and oceanic – differ in thickness, density and overall composition. The continents are the oldest part of the crust and therefore the lightest. This allows them to float higher up on the mantle than the oceanic crust – on average 300m (1,000 ft) above the present sea level.

Continental crust averages 30–40 km (18–24 miles) thick compared with 5 km (3 miles) for ocean crust.

Mount Villarica
This volcano in central Chile has erupted at least 20 times since 1600. It is located in the Andean subduction zone.

Ocean Crust

The ocean floors are much younger than the continents. Oceanic crust is recycled as it is extruded at ocean ridges and pushed back into the mantle at subduction zones. The Earth's oldest ocean floor was formed 200 million years ago (the oldest continental crust is over 3,000 million years old). Oceanic crust is denser than the continents; it is also thinner, on average only 5 km, (3 miles) thick, and lies on average 500 m (1,640 ft) below current sea level. Flat abyssal plains cover much of the ocean floors, interrupted by ridges and trenches.

Cratons

The ancient stable centres of the continents are known as *cratons*. They are generally flat and lie near sea level. They contain the oldest rocks of the crust. Disturbance of the cratons now tends to be by gentle warping that produces arches and basins a few hundred kilometres in diameter. Though now stable, cratons were also formed by turbulent geological events.They often comprise two or more ancient pieces of continent, which have been joined.

Ocean Ridges

Submarine ocean ridges can reach up to 3,000 m (10,000 ft) in height, towering over the surrounding abyssal plains. Mantle convection currents rise at these ridges and then flow horizontally in opposite directions. Vast amounts of volcanic material are discharged from the ridges, creating the ocean floors. Ocean ridges are long and volcanically active, registering heat flow from the interior with many shallow focus earthquakes.

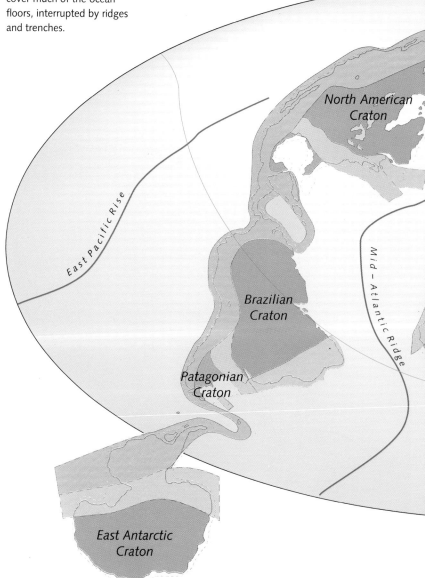

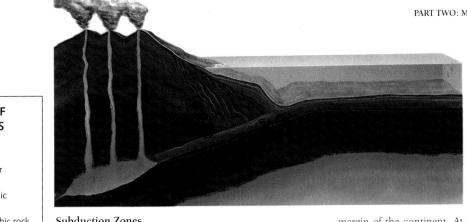

EPICENTRES OF EARTHQUAKES

- ocean
- ocean floor sediment
- cold oceanic plate
- metamorphic rock accreted onto continent and altered
- continental crust
- zone of melting, magma movements
- earthquake foci

Subduction Zones

Where two plates of the Earth's crust are driven towards each other, one will be pushed under the edge of the other and down into the molten mantle. Because it is made of heavier material, an oceanic plate will always be driven below a continental plate. The places where this happens are called subduction zones and are regions where most of the Earth's geological activity occurs.

As the ocean crust is forced downwards it heats up. Some of the resulting molten magma finds its way up through the continental crust and onto the surface, resulting in volcanoes and lava flows. As the volcanic material builds up into mountains it is rapidly eroded and huge quantities of sediment are dumped into the ocean trench that runs along the margin of the continent. At the same time sediments that have gradually accumulated on the ocean floor are pushed against the edge of the continent by the movement of the plates. All of these processes add material to the continental margin in the area of the subduction zone – it is here that the continents are growing. Subduction zones are also the sites of most of the world's earthquakes. As the two plates rub together the resulting friction causes juddering movements in the lower crust which are felt on the surface as earthquakes.

Mount St Helens
The eruptions of 1980 spread ash and dust over several hundred square kilometres. Eruptions in this region are caused by movements of plates in the Pacific, relative to North America.

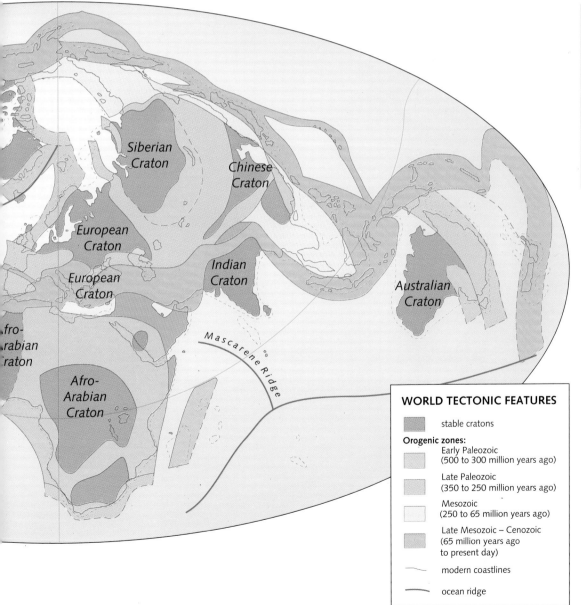

Siberian Craton

Chinese Craton

European Craton

European Craton

Indian Craton

Australian Craton

Afro-Arabian Craton

Afro-Arabian Craton

Mascarene Ridge

WORLD TECTONIC FEATURES

- stable cratons

Orogenic zones:
- Early Paleozoic (500 to 300 million years ago)
- Late Paleozoic (350 to 250 million years ago)
- Mesozoic (250 to 65 million years ago)
- Late Mesozoic – Cenozoic (65 million years ago to present day)
- modern coastlines
- ocean ridge

Lava Flow, Hawaii
The world's biggest volcanoes are on Hawaii which sits above a 'hot plume' in the mantle.

Atmosphere and Oceans
Earth's life support system

The formation of the Earth's unique atmosphere was a critical factor in the ability of the planet to sustain life. The most unusual aspect of the Earth's atmosphere is that it is constituted of nearly 20 per cent free oxygen – of the other planets only Mars has any free oxygen. The presence of free oxygen at least 3,500 million years ago is attested by the appearance of blue-green algae.

There are two main theories for the formation of the atmosphere and sea water. The theory of outgassing proposes that most atmospheric gasses have their origin in the interior of the Earth and are brought to the surface in volcanic and igneous processes. This explains the presence of nitrogen, carbon dioxide, helium, argon and water vapour in the atmosphere. In addition the steam expelled by volcanoes would be a ready source of sea water.

The theory of photochemical dissociation shows how an early atmosphere made up of water, methane and ammonia (like Jupiter's today), evolved into our current atmosphere, comprised mainly of nitrogen, oxygen, carbon dioxide and water vapour. The essential starting point is the action of sunlight on water, which splits it into hydrogen and oxygen for this process.

The answer is probably a combination of both processes. Once oxygen was present in small quantities life as we know it could begin and the process of photosynthesis could then produce more oxygen. No other planet in the solar system has the same type of atmosphere as Earth.

Hurricane Gladys *(right)*
This view from space shows the clockwise rotation of a southern hemisphere hurricane. Associated winds can reach 320 km per hour (200 mph). Hurricanes require a particular set of circumstances, but where these occur, hurricanes become a regular occurrence. A sea surface temperature of 27°C is a minimum requirement, restricting hurricanes to tropical regions.

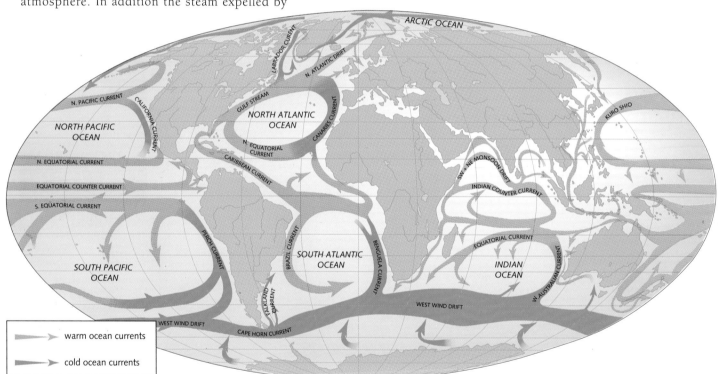

ARCTIC OCEAN

LABRADOR CURRENT

N. ATLANTIC DRIFT

N. PACIFIC CURRENT

CALIFORNIA CURRENT

GULF STREAM

NORTH ATLANTIC OCEAN

CANARIES CURRENT

KURO SHIO

NORTH PACIFIC OCEAN

N. EQUATORIAL CURRENT

N. EQUATORIAL CURRENT

CARIBBEAN CURRENT

SW & NE MONSOON DRIFT

EQUATORIAL COUNTER CURRENT

INDIAN COUNTER CURRENT

S. EQUATORIAL CURRENT

PERU CURRENT

BRAZIL CURRENT

SOUTH ATLANTIC OCEAN

BENGUELA CURRENT

EQUATORIAL CURRENT

SOUTH PACIFIC OCEAN

INDIAN OCEAN

W AUSTRALIAN CURRENT

FALKLAND CURRENT

WEST WIND DRIFT

WEST WIND DRIFT

CAPE HORN CURRENT

→ warm ocean currents

→ cold ocean currents

Ocean currents, landmasses and climate
The Earth's climate is profoundly affected by both deep water and surface ocean currents. The direction of the ocean currents is in turn affected by the circulation patterns in the atmosphere and by the positions of the continents.

Climate is affected because the ocean currents are constantly redistributing a huge reservoir of warm and cold water around the world. At some stages in the Earth's history the position of the continents has allowed currents to bring warm water to the polar regions and carry cold water away – giving a generally warmer and more evenly balanced climate across the Earth. But at other times, like our own, the poles are isolated by other landmasses and stay persistently cold. They develop ice caps, the whole Earth cools, and temperature differences between the poles and the equator increase.

Kilauea volcano
This eruption of Mount Kilauea on Hawaii, shows the vast quantities of steam that are produced by volcanic activity. Outgassing is the probable source of atmospheric gases and sea water.

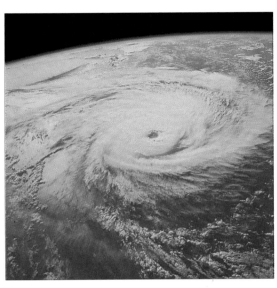

Earth without land
Oceans existed 4,000 million years ago, when the continents were insignificant. The prevailing winds and surface currents would have been as shown on this idealized globe *(above),* as straight lines influenced by the Earth's rotation *(black arrows).*

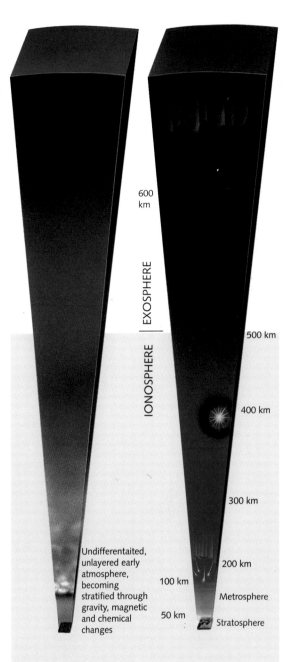

600 km

EXOSPHERE

IONOSPHERE

500 km

400 km

300 km

200 km

Undifferentaited, unlayered early atmosphere, becoming stratified through gravity, magnetic and chemical changes

100 km

Metrosphere

50 km

Stratosphere

Archean atmosphere

Present–day atmosphere

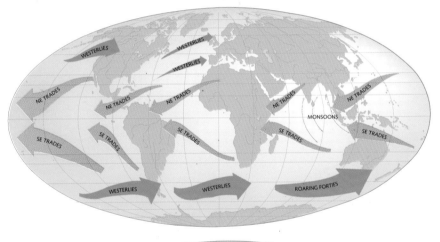

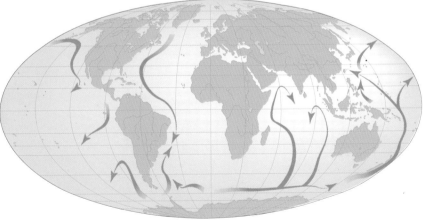

Deep currents *(above)*
When ocean water cools at the poles it becomes denser and sinks. With the rotation of the Earth, this water moves away from the polar regions towards the equator, stabilizing the Earth's climate.

Prevailing winds *(top)*
Atmospheric circulation is affected by the rotation of the Earth and the varying temperature of the Earth's surface. Prevailing winds create ocean currents by pulling the surface of the oceans around with them.

The Early atmosphere
The composition of the early atmosphere is crucial to understanding the development of the Earth, and is the subject of continuing research and dispute among scientists. The model shown here is therefore an appoximation.

The atmosphere is now more layered than it was. Oxygen and nitrogen molecules remain closely associated up to an altitude of 80 km (50 miles). Above that the heavier nitrogen sinks. Higher still hydrogen and helium predominate.

The Archean World
The Earth 4,000 to 2,500 million years ago

The Archean period is the oldest division of the geological time-scale. It begins with the creation of the first permanent rocks on the Earth's surface around 3,900 million years ago, and ends 1,500 million years later. It thus covers over a third of the the Earth's history.

At the beginning of the Archean the world had no atmosphere, no permanent land, no definable oceans, no recognizable life forms. By the end of the period small pieces of continental crust had established themselves permanently on the surface, an atmosphere containing some oxygen had formed, and there was abundant sea water. Single-celled life had taken a definite toe-hold on the planet showing a surprising diversity by 2,500 million years ago. Archean rocks are the oldest surviving parts of the crust—this is our starting point for the geological history of the Earth.

The Archean is part of the Precambrian era – all that part of the Earth's history before the start of the Cambrian period 550 million years ago. The Precambrian therefore covers about 90 per cent of the geological time-scale – though because it is the oldest period, fewer Precambrian rocks survive on the surface of the Earth than those of later periods.

Early geologists believed that Precambrian rocks predated the development of life on Earth. The Precambrian was left as a somewhat mysterious past, beyond the reach of the new science. It is now known that certain life forms existed quite early in the Precambrian and had become highly developed by the end of the period. But even with radioactive dating, electron microscopy and other modern techniques, the study of the Precambrian period, and particularly the Archean, is a matter of educated guesswork and speculation.

Archean provinces – the world's oldest rocks
Archean rocks occur on all the world's continents, emphasizing their role as the first stage in continental formation. There are undoubtedly larger areas of Archean rock underlying the more recent outcrops than appear on the surface. Archean rocks have been altered so much during their history that they offer little information about the geography of the Earth at the time. Any paleographic reconstruction before the Cambrian period is therefore highly speculative.

Greenstone belts
Archean greenstones occur within Archean provinces – the remnants of the first continents on Earth. These tiny slivers of land were formed in series in different places on the Earth's surface. Each micro-continent remained separate for some time, but then continental drift brought some of them together. Archean greenstones show the same 'belt' patterns in different parts of the world – most notably the Yilgarra province of western Australia (*far right*) and Superior province in eastern Canada (*below*). These most ancient rocks are rich in iron and other ores.

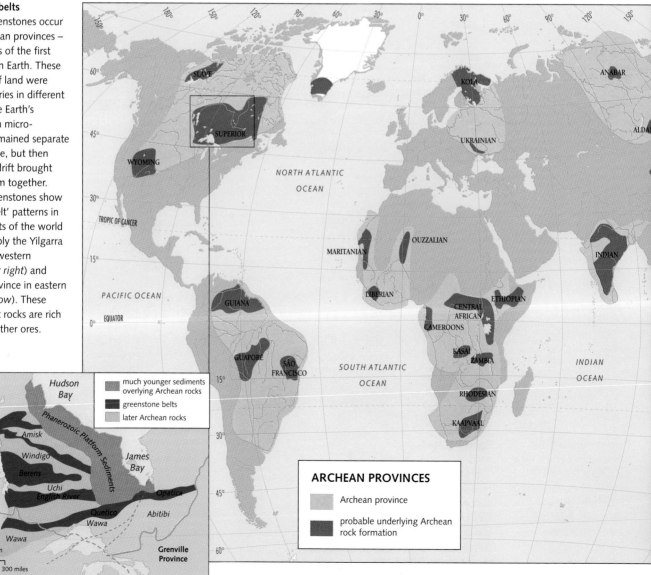

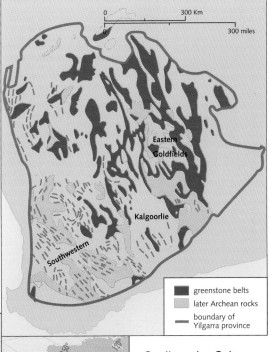

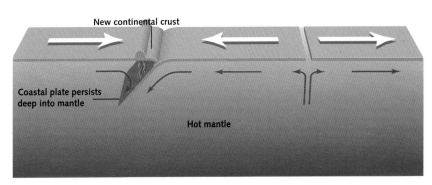

New continental crust

Coastal plate persists
deep into mantle

Hot mantle

greenstone belts
later Archean rocks
boundary of
Yilgarra province

Formation of the first continents

As the surface of the Earth cooled the plates formed on the surface. These were dragged down into the hot molten mantle on convection currents, which prevented permanent crusts forming. About 4,000 million years ago the plates became cold enough to bring about a change in the process of constant recycling of the crust. Without this crucial change the continents would never have formed, as has happened on the Moon.

The colder plates stayed solid deeper into the mantle as they were pulled downwards, and were subjected to greater heat than before. This produced a lighter material which floated to the surface and formed a different type of plate. Initially the plates were narrow strips reflecting the shapes of the fissures in the crust. Being lighter these plates have resisted being pulled back into the mantle and have stayed on the surface – as continents.

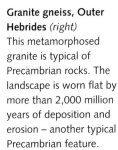

Granite gneiss, Outer Hebrides (right)

This metamorphosed granite is typical of Precambrian rocks. The landscape is worn flat by more than 2,000 million years of deposition and erosion – another typical Precambrian feature.

Evidence of Archean life

The oldest rocks on Earth contain traces of organic carbon – though this may not be evidence of life. Algal structures known as stromatoliltes have been found in rocks 3,550 million years old, in Warramoona, Western Australia, and bacterial structures are present in rocks 3,350 million years old in Swaziland. The search for older remnants of life continues – but is constrained by the formation of the first permanent continents.

Hornblende gneiss, Shetland (right)

The bands in this Archean rock were formed by low temperature metamorphic changes over a long period. They were originally horizontal bands.

The Origins of Life
A chance combination

Discovering more about the origins of life on Earth increases our sense of wonder at the emergence of such a phenomenon. The development of all but the simplest life forms involved the use of DNA, which not only has the necessary ability to store information to pass to future generations, but also has the crucial property of making occasional errors. Without these errors, or mutations, there would be no evolution, and life on Earth would have been unable to develop.

It is impossible to say exactly where and when life began on the Earth, but we know a good deal about the processes that were involved, and we can relate them to the conditions that existed on Earth at the time. The first stage in the development of life must be the formation of complex chemicals called amino acids, which are the building blocks of proteins, and can be be made from simple constituents under the right conditions. The necessary conditions include a reasonably high temperature and a high level of energy. There would certainly have been many places on Earth with high enough temperatures – pools of water near lava flows, hydrothermal vents, or even shallow tropical seas. The necessary energy would have come from ultra-violet light from the Sun, reaching the Earth without being absorbed, as there was no ozone layer.

But to have that seemingly mysterious property of life, a substance must be able to reproduce itself. The ability to reproduce means being able to pass on information to offspring, and this means developing a system that will carry that information. The system of information transfer which has developed on Earth is DNA (deoxyribonucleic acid). The vital stage in the origin of life on Earth is therefore the manufacture of DNA from proteins. DNA is an extremely complex substance which is not likely to have arisen spontaneously from much simpler proteins. It is thought that there were other less complex information systems around before DNA, but these systems were not as effective and life forms that depended on them died out when DNA-based life forms emerged.

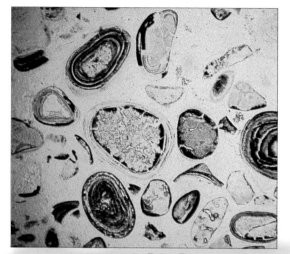

Gunflint Chert microfossils
This 2,000 million-year-old formation is a notable source of microfossils. Algal organisms have been identified *(left* and *right)*. Some of these were probably free-floating, while some formed colonies. The essential ability to photosynthesize was shared by all these early life-forms.

4,000 million years ago
Earliest life may have begun at around this time, though traces of carbon in the very oldest rocks may not be evidence of life.

3,500 million years ago
Bacterial structures developed – single-celled organisms have been found in rocks of this age in different parts of the world.

2,500 million years ago
Blue-green algae helped in the development of an atmosphere containing free oxygen.

2,000 million years ago
Single-celled organisms formed colonies of organic walled microfossils – predecessors of multi-celled organisms.

1,000 million years ago
Eukaryotic cells developed. These contain strands of genetic information, allowing sexual reproduction to take place.

ASIA

INDIAN
OCEAN

EVIDENCE FOR LIFE'S ORIGINS:
The Earth's oldest fossils

☐ Archean rocks

☐ Precambrian rocks

● notable Precambrian fossil
sites

Warrawoona

Pilbara ● Bitter Springs

● Ediacara

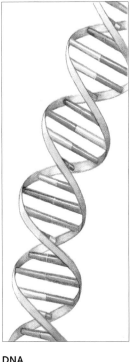

Black smoker
It had been thought that the earliest life must have developed in shallow water with access to sunlight for photosynthesis. But research into black smokers (photo, above) has shown that a rich variety of life-forms exist at great depths. These vents pour out hot sulphur-rich water from deep in the crust. The chemical mixture around black smokers creates an environment to which various life forms have adapted.

DNA
DNA is a highly complex molecule which has the ability to store information. The information is encoded by variations in the order in which elements of the molecule are stored on the double helix structure. Every cell in every organism contains DNA, and it is the DNA which instructs the cell in the order in which it produces vital proteins. DNA controls the cell's chemistry, including the synthesis of new DNA.

The origins of DNA itself are unknown. Such a highly complex molecule probably had less sophisticated predecessors as information systems – and these were replaced as DNA-based life forms took over.

Evidence for life's origins: the Earth's oldest fossils
Several places compete for the distinction of having the oldest fossils on Earth. The map (left) shows the earliest Archean sites presently known, and some of those known from later in the Precambrian. As techniques improve, still older microfsssils are found. But it is safe to say that the site of the origin of life on Earth will never be known.

Carbon found in the earliest rocks is described as 'organic' but this refers to its chemical form, rather than giving evidence of the presence of life. Rocks at Warrawoona in Northwest Australia are, at 3,500 million years old, among the oldest sedimentary rocks on Earth. They contain bacteria and blue-green algae. Precambrian rocks around 2,500 to 1,500 million years old have shown a surprising degree of development – including those of the Gunflint Chert in Ontario.

1,000–600 million years ago
Multicellular organisms developed. Complex soft-bodied animals are preserved in a few places.

500 million years ago
Marine animals developed shells and skeletons which are easily fossilized and preserved in vast numbers in the geological record.

400 million years ago
The first life on land. Early vascular plants were followed onto land by amphibians and insects. reptiles later developed from amphibians.

200 million years ago
The evolution of the first mammals, from the therapsid reptiles. Dinosaurs dominated until 65mya, mammals and birds have since taken over.

4 million years ago
The evolution of the first hominids. The earliest known hominid fossil is just under 4 million years old, though hominids separated from the great apes around 10 million years ago.

The Making of Rocks
Solid rock from sediment

All the rocks on the Earth are of volcanic or igneous origin. They were formed from molten magma that originates in the mantle. Magma continues to pour onto the Earth's surface through volcanoes and fissures in the crust. But the rocks of the Earth are continually eroded and reformed, and it is this process that leads to the deposition of *sedimentary rocks*.

There are a huge variety of sedimentary rock types, reflecting the range of materials from which they are formed and the environments in which they were deposited, as well as the conditions that affected them after deposition. It has been the task of geologists over the last two centuries to unravel the clues that sedimentary rocks offer about past environments. If we want to know what the world was like at particular times in the past, sedimentary rocks are the first place to look. Sedimentary rocks are also the medium for the preservation of fossils – the direct evidence of past life forms.

Physical and chemical sedimentation

The deposition of sediment on the Earth is generally the settling out of solid matter from water, the only important exception being the formation of desert sandstones. Solids settle out of suspension, either because of gravity (known as physical sedimentation), or by balancing the composition of seawater by the precipitation of dissolved minerals (chemical sedimentation). Physical sedimentation produces sandstones and shales, which are comprised of pieces of insoluble material washed off the Earth's surface by rain, rivers and waves. These are known as clastic sediments. Chemical sedimentation produces limestones and chalks that are comprised of soluble carbonate minerals which are precipitated out of large bodies of water.

Fast rushing streams carry heavy boulders

Tranquil lakes deposit fine silt

Water and sediment

Clastic sediments constitute over three-quarters of the mass of sediments on the Earth's crust. They are formed by a process of erosion, transportation and deposition. This process works downhill, and the deposition of sediments occurs at the lowest possible point. Smaller grains of sediment, which stay in suspension in water much more easily, are taken further and deposited in more tranquil situations, such as deltas. Large rocks and pebbles are deposited near to the site of erosion.

Boulders and conglomerates

Fast-rushing mountain streams are able to carry large boulders over short distances (*right*). Smaller pebbles and fragments are more likely to be buried and compacted into rocks known as conglomerates (*far right*). Sedimentary rocks are formed when sediments are buried and squeezed by overlying deposits. The pressure brings about physical and chemical changes which harden the sediment – the process is known as lithification.

Sandstones

Larger grained rocks can tell us more about the environment of deposition, because of their closeness to their point of origin. Sandstones give ready clues as to their origins. Geologists study the grain size, how rounded grains are, and how well sorted they are. Well-rounded grains, for example, will indicate a long journey before deposition. Geologists also study markings on the sandstone strata such as cross bedding. This is an indication of the prevailing current of water at the time of deposition. Sandstones sometimes carry ripple marks, just like the sand on a beach, which are another indication of current movements. Sandstones are usually deposited in coastal and delta conditions.

Gravel, sand and silt are deposited in bends and meanders

Shales and mudstones

As the river reaches its flood plain only fine mud and silts are held in suspension by the slow-moving water. They will be dropped in tranquil places like the inside of bends, or at deltas, and this fact offers geologists a clue in reconstructing environments. Mud and shale are the most common sediments, but are also the most difficult to study. The grains can only be seen by using electron microscopes. Slates (*below*) are finely cleaved mudstones.

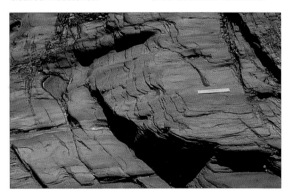

Beach sediment

Sediments are formed, washed away, and reformed continually. At every beach in the world each tide brings more material and washes some away. Only a tiny proportion of the material deposited is buried and becomes rock. Sandstones formed on beaches sometimes show ripple marks (*right*) – like those we see on present-day beaches. They are an indicator of ancient tide directions, which in turn show coastline variations.

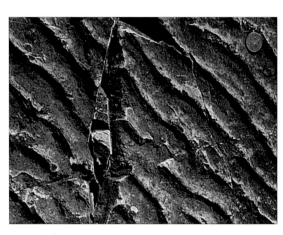

Ocean sediments

In the deep oceans, sediments accumulate very slowly, but are frequently preserved because of the relatively undisturbed conditions. In shallow waters around coasts, estuaries and on continental shelves, deposition is much more rapid but much more temporary. In shallow water wave action limits the depth of deposit.

The most likely environments for rapid, continuous deposition are places where depressions in the sea bed occur alongside new mountain chains. Massive amounts of sediment can accumulate in these depressions, which are known as *geosynclines*. The process is exaggerated by the fact that the weight of sediment pushes the crust down, allowing even more sediment to accumulate on top. A present day example is the Peru–Chile trench, which runs alongside the Andes Mountains.

Satellite image of the Ganges Delta

The Himalayas, (*right*) one of the Earth's newest mountain ranges, are being eroded very quickly. A vast amount of eroded material is washed off the mountains and carried down into the Ganges basin by fast flowing streams. The finer material is carried further; some is deposited at the mouth of the Ganges in the network of deltas. The rest is carried a little way out to sea, where a huge bank of sediment, known as the Ganges Cone, has built up on the sea floor.

Huge quantities of sand and silt are deposited in estuaries and deltas

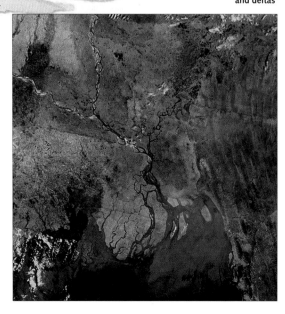

The Proterozoic World
The Earth 2,500 to 550 million years ago

The Proterozoic period saw the establishment of the continents of the Earth. From about 2,500 million years ago the land areas were transformed from narrow strips of crust into continent-sized pieces. These ancient hearts of the continents are known as cratons. During the past 2,500 million years these cratons have gradually grown into the continents that we see today. The low-lying eroded remains of these ancient continents can be seen today in the ancient shields of Canada, Siberia, Australia, central Africa, the Amazon and Antarctica.

Proterozoic literally means 'early life', and the period occupies the latter part of the Precambrian era. Having discovered that there was life in the Precambrian period after all, geologists dated its beginnings to the start of the Proterozoic. But once again scientific methods have improved, and evidence of life has been found in rocks as old as 3,500 million years.

The long time-span of the Proterozic was undoubtedly a crucial period in the development of the planet. At the beginning, small pieces of land sat in the middle of vast oceans. An atmosphere with little oxygen supported a few fragile life-forms. By the end of the Proterozoic era, the atmosphere was rich in oxygen and marine life was plentiful – ready for the spectacular developments of the next 500 million years.

While all the continents of the Earth have

centres, or cratons, made of rocks largely of Proterozoic age, it is very difficult to determine the position of the continental cratons at such an early time in their history. There is a theory that, at some time in the late Precambrian, there was an early super-continent, or Pangea, which was made up of all the continents. A smaller Precambrian supercontinent may have comprised Laurentia, Baltica and Siberia.

Landscape of the Canadian Shield (below)
Younger sediments have been scraped off the shield, revealing a flat landscape of Precambrian rock.

The Canadian Shield
Continental cratons, the stable centres of the Earth's continents, are complex structures, formed in a series of catastrophic events over many millions of years. The most studied of all is the Canadian Shield – a vast area of Precambrian rock covering the eastern two-thirds of Canada, extending south to Lake Superior and east to western Greenland. It is 5 million square kilometres, with a shield-like shape on the map, and a flat, slightly domed topography.

This low-lying area is surrounded by more recent mountainous belts. Since its formation, the great mountain chains of the Canadian Shield have been eroded away, as have any younger rocks deposited on top of the Precambrian basement.

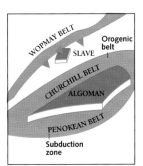

2,000 million years ago
The Algoman and Slave provinces were separate continental blocks. They were formed from cooling of parts of the upper mantle, then drifted towards each other, creating a subduction zone, the Churchill province. This was the site of major mountain-building, volcanism and outpourings of granitic lava.

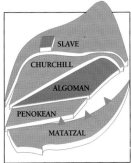

1,500 million years ago
The continents of Siberia and Baltica were fused to North America around 1,500–1,200 million years ago, causing volcanism and mountain-building on both. The Slave and Algoman provinces were surrounded by newer continental crust. The continent of Matatzal converged from the south c.1,300 million years ago.

1,200 million years ago
Matatzal collided with the new continent in the south, remaining attached along the whole of its northern margin for several million years. Evidence of outpourings of basalt lava indicate the rifting away of most of Matatzal province. A portion was left behind, now underlying the southern Great Plains.

1,000 million years ago
The Grenville belt was formed along the south-east of the continent – presumably by the fusion of another continental mass. This completion of the craton left a mosaic of old orogenic zones – now fused into a stable structure. It is overlain in places by younger sediments – though in Canada many of

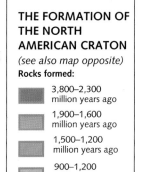

THE FORMATION OF THE NORTH AMERICAN CRATON
(see also map opposite)
Rocks formed:

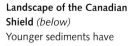

3,800–2,300 million years ago

1,900–1,600 million years ago

1,500–1,200 million years ago

900–1,200 million years ago

400 million years ago to present day

these were scraped off by the Pleistocene ice-sheets. Subsequent events have added continental crust to the margins of the old craton – most notably the Appalachian and Cordilleran orogenic episodes.

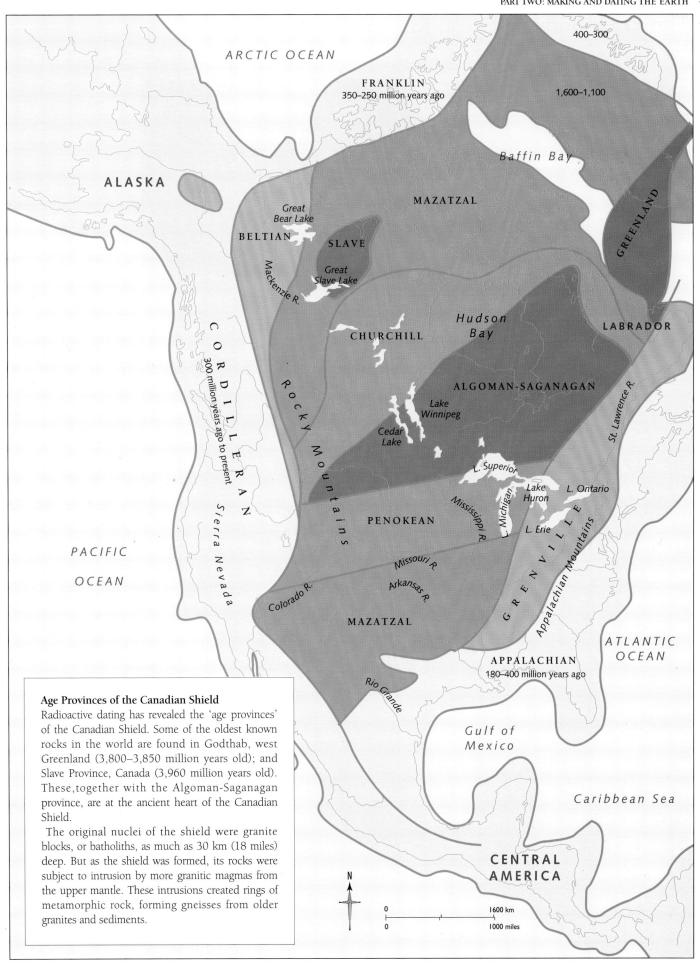

ARCTIC OCEAN

FRANKLIN
350–250 million years ago

400–300

1,600–1,100

Baffin Bay

ALASKA

Great
Bear Lake

BELTIAN

SLAVE

Great
Slave Lake

GREENLAND

Mackenzie R.

CHURCHILL

*Hudson
Bay*

LABRADOR

ALGOMAN-SAGANAGAN

St. Lawrence R.

Lake
Winnipeg

Rocky Mountains

C O R D I L L E R A N
300 million years ago to present

Sierra Nevada

Cedar
Lake

L. Superior

Lake
Huron

L. Ontario

PACIFIC

OCEAN

Mississippi R.

L. Michigan

L. Erie

PENOKEAN

G R E N V I L L E

Appalachian Mountains

Missouri R.

Arkansas R.

Colorado R.

ATLANTIC
OCEAN

MAZATZAL

MAZATZAL

APPALACHIAN
180–400 million years ago

Rio Grande

*Gulf of
Mexico*

Caribbean Sea

Age Provinces of the Canadian Shield
Radioactive dating has revealed the 'age provinces'
of the Canadian Shield. Some of the oldest known
rocks in the world are found in Godthab, west
Greenland (3,800–3,850 million years old); and
Slave Province, Canada (3,960 million years old).
These, together with the Algoman-Saganagan
province, are at the ancient heart of the Canadian
Shield.
 The original nuclei of the shield were granite
blocks, or batholiths, as much as 30 km (18 miles)
deep. But as the shield was formed, its rocks were
subject to intrusion by more granitic magmas from
the upper mantle. These intrusions created rings of
metamorphic rock, forming gneisses from older
granites and sediments.

CENTRAL
AMERICA

N

0   1600 km

0   1000 miles

Early Life and the Fossil Record
Discovering the past in the present

We now take it for granted that fossils preserved in rocks are the remains of creatures, often extinct, from the distant past. This understanding was central, not only to the foundation of geology in the early 19th century, but to the developments of Darwin's ideas on evolution.

Before then it had been suggested that the close relations of the exotic creatures whose bones had been discovered in quarries and mines were still living somewhere on Earth – there might be mammoths in deepest Africa, or dinosaurs in the unknown west of America. This idea was gradually abandoned as less and less of the world remained to be explored by Europeans. By the time Darwin went to South America in the 1830s he believed that the fossil animals he found were of extinct species – but that they bore a strong relation to the animals currently living in the region.

By that time most scientists had accepted that the age of the Earth was to be measured in millions, not thousands, of years. This revised estimate supported the idea that fossils were the record of the life-forms that had come and gone on the Earth, and allowed for the necessary time for the gradual changes in natural evolution to produce the diversity of life-forms that have existed on Earth. Once this was established paleontologists were able to use the fossil record to devise an intricate and general system of geological dating. Assessments of the relative ages of rocks were based on fossils until the additional use of radioactive dating.

In addition, fossils are invaluable indicators of ancient environmental conditions. These range from the crude limits placed on marine life by sea levels, to the exquisite and mysterious mechanism whereby the shells of tiny Foraminifera from the Pleistocene period coiled to the left when sea temperatures were below 9°C and to the right when they were higher.

Until recently it was assumed that life forms on Earth remained extremely simple until the beginning of the Cambrian period 550 million years ago. The discovery of fossils in the Ediacaran rock formation of Australia gave dramatic evidence that more complex life had developed somewhat earlier. Whereas most fossils are formed from shells and skeletons, the Ediacaran rocks have preserved the soft bodies of the plants and animals living there between 600 and 700 million years ago.

Ediacara jellyfish
This fossil jellyfish is from the Ediacara formation of southern Australia. The host rock is 600 to 700 million years old.

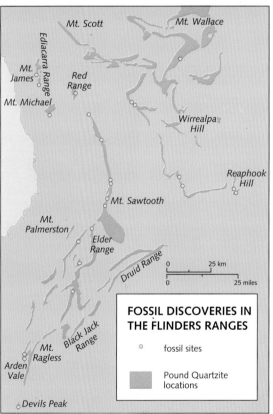

FINDS OF EDIACARA FOSSILS

Mt. Skinner
Jervois Ranges
Laura Creek
Deep Well
Punkerri Hills,
Officer Basin
see inset
Ediacara,
Flinders
Range

Mt. Scott
Mt. Wallace
Ediacarra Range
Mt. James
Red Range
Mt. Michael
Wirrealpa Hill
Reaphook Hill
Mt. Sawtooth
Mt. Palmerston
Elder Range
Druid Range
0   25 km
0   25 miles
Black Jack Range
Mt. Ragless
Arden Vale
Devils Peak

FOSSIL DISCOVERIES IN THE FLINDERS RANGES

○   fossil sites

Pound Quartzite locations

Early life forms in Ediacara
Deposits of Pound Quartzite in the Flinders Ranges of southern Australia (*left*) contain fossils of Ediacaran fauna – named after the location where they were first found. These 700 million-year-old plants and animals from Australia show some remarkable characteristics (*below*). They have developed the ability to consume food, rather than manufacture it themselves. They have also evolved specialized body parts for locomotion and a body cavity for feeding and digestion. They are mostly jellyfish of different kinds. Some were free floating, some were attached as 'sea fans'. Some resemble later groups like molluscs, but were probably unrelated to them.

Stromatolites

Once cellular life forms had emerged, the next stage was for single-celled organisms to join together. The most spectacular early examples of this colony formation are the stromatolites (*below*). These huge reef-like structures are formed by lime-secreting blue-green algae (cyanobacteria). These form 'mats' of lime, often with layers of mud from the sea bed trapped between, giving them a wavy, laminated structure which stands out when they are fossilized in rocks.

In Precambrian times the Moon was closer to the Earth, leading to bigger tides and greater depths of intertidal waters where stromatolites flourished. Fossil remains have been found in Precambrian rocks in Canada, South Africa and Australia (*map above*).

PRECAMBRIAN STROMATOLITES
• finds of stromatolites

Shark Bay, Australia
(*above*)
Stromatolites still exist in some parts of the world – most notably in Shark Bay, Australia. It is unususal to see such a number and size, since they are normally heavily 'grazed' by marine invertebrates.

Fossils and dating

The main basis for understanding the history of our planet is our ability to date accurately the rocks of the Earth. Fossils are an essential tool in dating rocks, and in cross-matching the ages of rock formations that are physically separated.

Fossils are also a direct record of the evolution of life on Earth. The fossil record is unfortunately far from complete, the preservation of fossils being a chancy process. Some plants and animals are much more easily preserved than others, and therefore more common in the fossil record. Certain environments are more conducive to preservation. The fossil record and individual fossils therefore need careful interpretation if they are to give an accurate picture of the past.

Sedimentary rocks of the same age are not all the same type. In one location limestone will have been formed, in another shale, and in another sandstone. But sometimes the same fossils will appear in almost all sedimentary rocks of the same age. These are known as *zone fossils* or *marker fossils*. Each geological period has a particularly useful zone fossil. The ideal zone fossil is a creature that existed for a short period of time and spread rapidly and widely. They then define a limited time zone over a large area.

Zone fossils, which are used to stratify geological periods (*right*), are usually rapidly evolving, free-swimming or floating creatures, which will spread widely over a comparatively short period. Trilobites, graptolites and brachiopods are the most useful zone fossils in the Paleozoic era (Cambrian to Permian periods). The dominant zone fossils of the Mesozoic (Triassic to Cretaceous) are the ammonites. In the Tertiary period foraminifera are used, and the beginning of the Quaternary, 2 million years ago, is marked by the presence of fossilized teeth and bones of horses, bears and elephants. In some periods such as the Triassic zone fossils are scarce, due to lack of marine sediments. This makes the history of these periods particularly difficult to unravel.

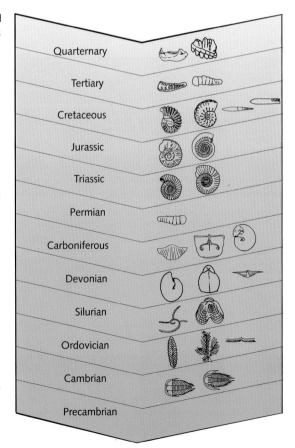

Early Glaciations
The Earth cools and freezes

Evidence of ancient ice-caps in tropical Africa was one of the most astonishing and perplexing discoveries of pioneering geologists. Glacial deposits known as tillites, and characteristic striations on rocks were identified in places where the climate was anything but icy. How could this evidence of ice-sheets at the tropics be explained? Either the whole world had been encased in ice at some time in the past, or the tropical continents had previously been located at the poles.

Evidence of glaciation has now been found on every continent on Earth. By piecing this together with other evidence of continental movement, we can build up a history of the ice ages that have affected the Earth throughout its history.

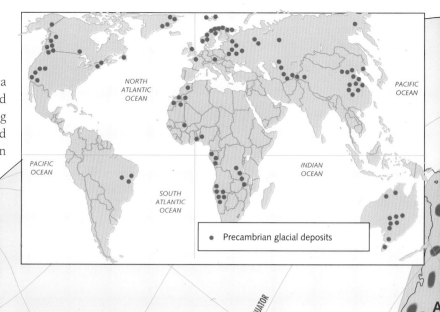

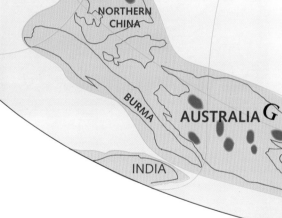

Evidence of glaciation

Because parts of the Earth are presently covered in ice-sheets and glaciers we have a ready-made demonstration of the rock types that are formed during glaciation. In Ontario in southern Canada there is a peculiar rock formation that is made up of a mixture of huge boulders, pebbles, clay and sand. Combined with this are fine-layered mudstones, some containing scattered pebbles. Where the rocks are exposed it is possible to see striations – long scratches or gouges out of the rock. The rocks are extremely ancient – about 2,200 million years old.

Exactly the same types of rocks are being created today 1,600 kilometres (1000 miles) further north. As the glaciers and ice-sheets of Greenland, Alaska and the north Canada coast meet their coastal waters, they deposit a huge quantity of mud and sand containing boulders and rocks of every size. At the margins of the ice-sheets glacial lakes form, where the mud is deposited at the bottom in seasonal layers – giving rise to the fine laminated mudstone seen in Ontario. From this comparison, we can deduce that the climate of southern Canada 2,200 million years ago was similar to that of the Canadian Arctic today.

The Earth's first ice ages

The first ice age occurred about 2,300 million years ago. Tillites from this ice age have been found in North America, Australia and southern Africa. Later in the Precambrian there is more widespread evidence of a major glacial episode that affected every continent on Earth. Tillites and other glacial deposits were formed across a very wide area. When the position of the continents 700 million years ago is mapped (*above*) it appears that glaciation occurred at every latitude from the poles to within 11° of the equator. It is possible that, during this 100-million-year ice age, the continents moved in and out of the polar regions, becoming glaciated in turn. It has also been suggested that the orbit of the Earth was exceptionally elliptical at this time. This led to extremes of temperature as the Earth orbited nearer and further from the Sun than at pesent. Studies of subsequent cold periods in the Earth's history – including the present – suggest that the configuration of the continents might be a major factor in cooling the climate. Land masses may prevent warm air and water from reaching polar regions. This may be enough to trigger the formation of ice sheets.

Striations on basalt

At some time in the past glaciers have polished and scratched the tops of these basalt columns at the Devil's Post Pile in California (*above left*) – now a region of warm climate. Parallel striations like these are only caused by hard objects frozen into the bottom surface of ice-sheets, and are therefore a definite indication of past glaciation. The scratches reflect the direction of ancient ice-flows.

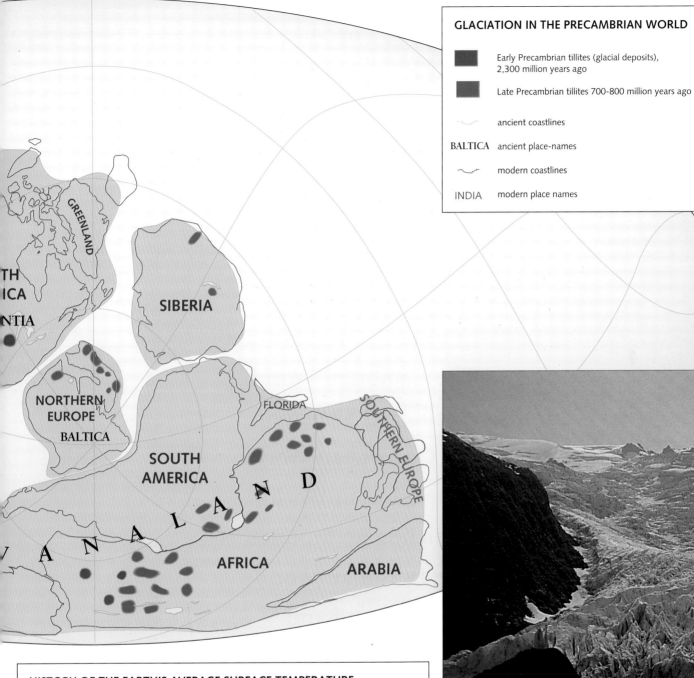

GLACIATION IN THE PRECAMBRIAN WORLD

Early Precambrian tillites (glacial deposits), 2,300 million years ago

Late Precambrian tillites 700-800 million years ago

ancient coastlines

BALTICA   ancient place-names

modern coastlines

INDIA   modern place names

GREENLAND

SIBERIA

NORTH AMERICA

LAURENTIA

NORTHERN EUROPE

BALTICA

SOUTH AMERICA

FLORIDA

SOUTHERN EUROPE

GONDWANALAND

AFRICA

ARABIA

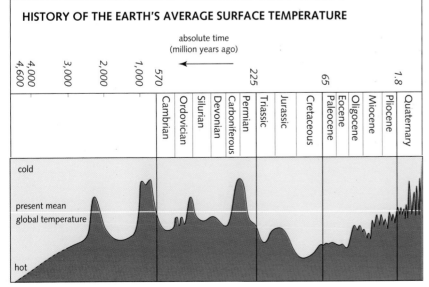

HISTORY OF THE EARTH'S AVERAGE SURFACE TEMPERATURE

absolute time
(million years ago)

4,600
4,000
3,000
2,000
1,000
570
225
65
1.8

Cambrian
Ordovician
Silurian
Devonian
Carboniferous
Permian
Triassic
Jurassic
Cretaceous
Paleocene
Eocene
Oligocene
Miocene
Pliocene
Quaternary

cold

present mean
global temperature

hot

Glacial episodes
The average temperature of the Earth's surface has varied through its history. A regular temperature cycle, combined with other factors such as ocean currents and the positions of the continents around the poles, all combine to influence the Earth's temperature.

South Holgate Glacier, Gulf of Alaska
Present-day glaciers, like this one in Alaska, show how frozen conditions radically alter the land surface. Ancient glaciers leave evidence of similar effects in the geological record.

Part Three
The Early and Middle Paleozoic Era

Evolutionary Explosion

The Cambrian World

Shells and Skeletons

The Ordovician World

The Silurian World

The Caledonian Mountains

The Devonian World

Life onto Land

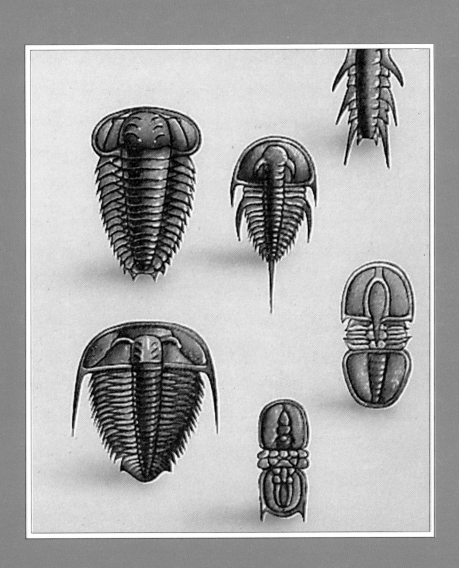

Evolutionary Explosion
Early marine life of the Burgess Shale

Our ideas about the development of life on Earth change as we discover more evidence of early fossils. The fossils of the Burgess Shale – exquisitely preserved, complex marine creatures – discovered in rocks as old as 530 million years, showed that there had been a massive development of marine life in a relatively short space of time. As multicellular organisms developed and spread it seems that there was rapid diversification of life forms. The environment was 'empty' and animals evolved quickly to take advantage of the lack of competition from other species. This was the first great evolutionary explosion.

The discovery of huge quantities of flattened but complete animals from a distant period, when animals lived only in the sea, sheds a uniquely brilliant light on one episode in Earth history. Not only are the animals complete, they are preserved in exquisite detail. In most cases, the antennae and limbs are preserved intact. In some instances, the body walls have been worn away and the internal structure of the animal is preserved. But it is only in the last twenty years or so that the real significance of the Burgess Shale fossils has come to be understood. The initial find in 1910 was spectacular and scientists around the world were amazed by the creatures and by the completeness of their preservation. Paleontologists assumed that the fossils represented primitive early forms of modern groups of animals. This fitted in with contemporary evolutionary thinking – that groups of organisms develop from a single primitive starting point and gradually become more 'advanced' with time.

But the new work on the Burgess Shale has shown that many of the fossils cannot be classified into modern groups. They represent distinct and separate groups of animals that died out relatively soon after the Burgess Shale was formed. This changes our view of evolution. Many of these animals were far from primitive, and were well adapted to their environment. The most startling fact is that there is no apparent reason why some groups survived. If we were able to travel back 530 million years and look at the sea bed community of animals, it would be impossible to forecast which would endure and which would die out. Whatever change took place in the Late Cambrian – change of sea level, chemical composition of the water, temperature – some creatures were lucky enough to be able to survive and flourish in the new environment, while others were not.

One of the creatures of the Burgess Shale, Pikaia gracilens, is the oldest known example of the group Chordata. This group comprises all the vertebrates including, of course, mammals and human beings.

There may have been other primitive chordates swimming in the Cambrian seas, but it is at least possible that our future existence depended on the ability of Pikaia to survive the decimation that followed the Burgess Shale. Given the number of Pikaia specimens found it seems that this was not one of the more abundantly thriving animals of the time. However, it survived, and is part of the reason why we are here.

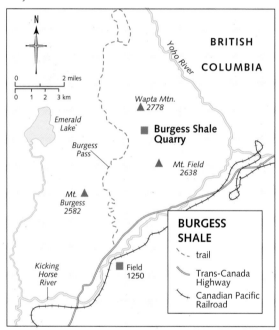

Burgess Shale sea floor
This area of the Canadian Rockies was on the western margin of the continent of Laurentia. There is a radical break in rock types in this area from coarse crystalline dolomite to shale and mudstone. It represents a huge ancient submarine cliff varying from 100 to 300 metres (300–900 ft) high along its 20 km (12 miles) length. The Burgess Shale was laid down at the foot of this gigantic underwater cliff.

OFF-SHORE REEF AND LAGOON IN THE CAMBRIAN PERIOD

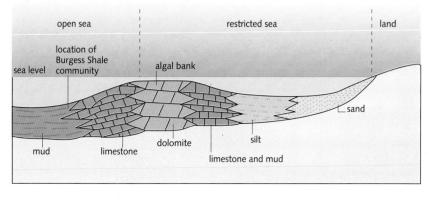

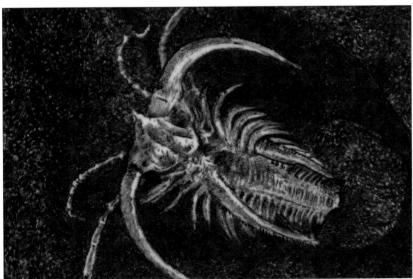

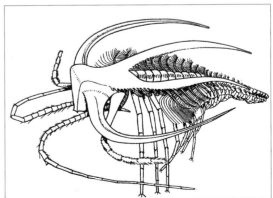

Marrella Splendens
This small arthropod was the first fossil found in the Burgess Shale. Its beautiful body structure has been reconstructed (*above*) from a series of fossil imprints found in the finely-grained mudstone.

The Cambrian World

The Earth 535 million to 500 million years ago

The Cambrian period is the first part of the era known as the Paleozoic. Much of the early work on the Paleozoic was done in Wales. This is reflected in the names of the geological periods. The ancient name for Wales is Cambria.

The start of the Cambrian period was first defined by geologists by the appearance of life-forms preserved as fossils. But we now know that the earliest life dates back to 2,500 million years *before* the start of the Cambrian. Cambrian rocks are now seen as the oldest in which fossils are sufficiently numerous and distinct, because of the presence of shells and skeletons, to provide reliable geological information.

Cambrian rocks are found on all continents. They are rich sources of zinc, lead, gold and silver. By the end of the Cambrian period a shallow sea, known as the Sauk Transgression, covered large areas of the ancient continents. As a result, most Cambrian formations contain significant amounts of carbonate rock (limestone and chalk).

The Cambrian World

The map of the Cambrian world is the first one in which significant pieces of our present continents can be plotted in their original positions. Coastlines on this map are averaged out for the period, and are in any case approximate. During the Cambrian era, most continents were in tropical latitudes, although the vast continent of Gondwanaland extended south to the polar regions. The continent of Laurentia was drifting apart from Baltica and Siberia, opening up a new ocean known as the Iapetus Ocean. This early version of the Atlantic continued to widen throughout the Early Paleozoic era.

Flora and Fauna

The diversity of life in the Cambrian seas was extraordinary, with many invertebrate groups making their first appearance. Even the most sophisticated Precambrian organisms are extremely simple compared to the plants and animals of the Early Cambrian. Some time around the start of the Cambrian period there was a crucial development in the evolution of marine life – animals developed the ability to manufacture shells and skeletons. This may be related to chemical changes in sea water, and the need for increased mobility.

The Cambrian fauna were predominantly marine creatures dwelling in shallow seas on the continental margins. There were no freshwater or land organisms in the Cambrian. Blue-green, green and red algae were the main marine plants, and these formed massive structures, called stromatolites. There are also traces of plants bearing strong resemblances to modern types of seaweed.

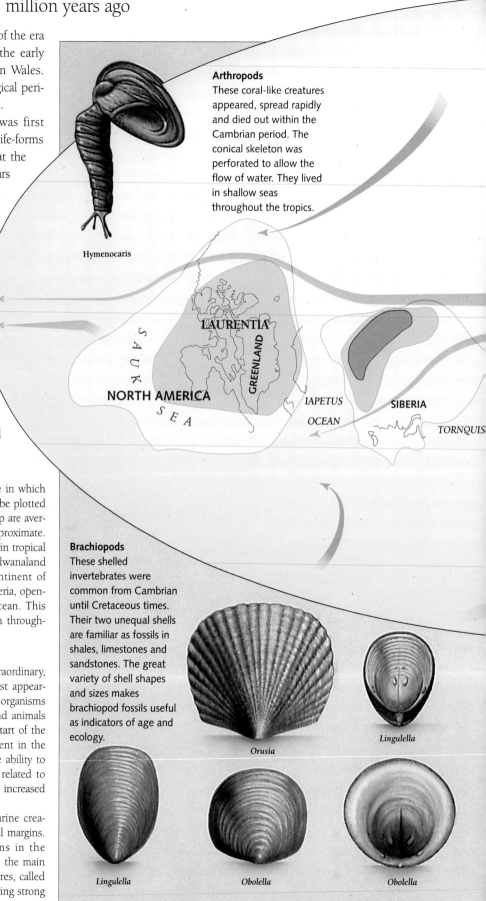

Hymenocaris

Arthropods
These coral-like creatures appeared, spread rapidly and died out within the Cambrian period. The conical skeleton was perforated to allow the flow of water. They lived in shallow seas throughout the tropics.

LAURENTIA

GREENLAND

SAUK SEA

NORTH AMERICA

IAPETUS OCEAN

SIBERIA

TORNQUIS

Brachiopods
These shelled invertebrates were common from Cambrian until Cretaceous times. Their two unequal shells are familiar as fossils in shales, limestones and sandstones. The great variety of shell shapes and sizes makes brachiopod fossils useful as indicators of age and ecology.

Orusia

Lingulella

Lingulella

Obolella

Obolella

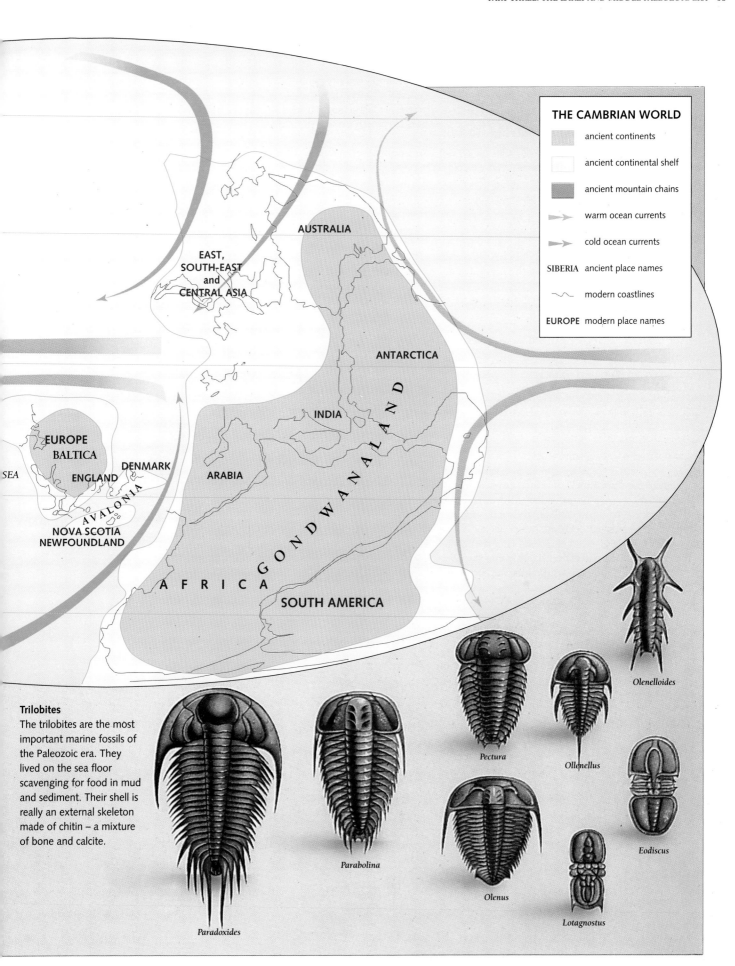

THE CAMBRIAN WORLD

- ancient continents
- ancient continental shelf
- ancient mountain chains
- warm ocean currents
- cold ocean currents
- SIBERIA ancient place names
- modern coastlines
- EUROPE modern place names

AUSTRALIA

EAST,
SOUTH-EAST
and
CENTRAL ASIA

ANTARCTICA

INDIA

EUROPE
BALTICA

DENMARK

ENGLAND

SEA

AVALONIA

ARABIA

NOVA SCOTIA
NEWFOUNDLAND

G O N D W A N A L A N D

A F R I C A

SOUTH AMERICA

Olenelloides

Pectura

Ollenellus

Trilobites
The trilobites are the most important marine fossils of the Paleozoic era. They lived on the sea floor scavenging for food in mud and sediment. Their shell is really an external skeleton made of chitin – a mixture of bone and calcite.

Eodiscus

Parabolina

Olenus

Lotagnostus

Paradoxides

Shells and Skeletons

Guides to the history of life on Earth

Most of the knowledge we have about the development of animal life on Earth comes from the shells and skeletons of animals preserved as fossils. The seemingly sudden appearance of animals with shells and internal skeletons (known generally as hard parts) at the beginning of the Cambrian period 535 million years ago is one of the major events in geological history.

Because most animals have a strict geographical range, fossils are a major clue to the positions of ancient seas and continents. Shells and skeletons are also part of the chemical balance of the Earth, taking massive amounts of calcium and other carbonates, phosphates and silica out of seawater in the course of their formation. Shelled animals formed a new type of sediment as their hard bodies were deposited on sea floors. They also made sructures such as reefs, which are preserved as rocks.

Early animals with shells and skeletons

Hard parts evolved in two ways; external skeletons, which appeared in four major groups – archeocyathids, brachiopods, molluscs and arthropods; and internal skeletons which appeared in echinoderms.

At the start of the Cambrian period, 535 million years ago, most of the main groups of marine invertebrate animals acquired the capacity to produce hard parts. Some animals produced hard parts of lime, phosphate or chitin – a horny organic material. Others had shells and skeletons made of mixed layers of chitin and lime, and chitin and phosphate. All the major invertebrate groups that have ever existed on Earth appear in the fossil record of the Cambrian period. Nearly all Cambrian life belongs to three groups; archeocyathids, brachiopods and arthropods.

Cambrian faunal provinces

A faunal province is an area in which a particular set of animals is found, and which is surrounded by areas with entirely different sets of animals. Using fossils from Cambrian rocks, geologists discovered that both the 'American' and 'European' faunal provinces crossed the 2,000-kilometre (1,250-mile) expanse of the North Atlantic (*maps below*). This gave strong evidence that the two continents were close together during the Cambrian period – as the reconstruction shows.

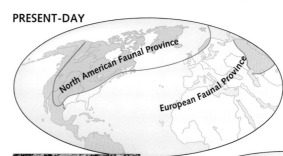

PRESENT-DAY

CAMBRIAN CONTINENTS

Calymene
Trilobite fossil (*above*) found in Dudley, England. Known locally as the 'Dudley Locust'.

Lloydolithus
Early Paleozoic trilobite, (*left*), one of the earliest animals to develop an external skeleton.

Archeocyathids

This group of animals, extinct by the middle Cambrian, was the first to secrete a hard, calcareous skeleton. They were hollow, cone-shaped animals, with inner and outer calcareous walls, which were porous and joined by radial septa. They lived in large numbers, attached to the sea bed parallel to the coastlines of shallow Cambrian seas.

Pycnoidocyatheus Profundus

Molluscs

Molluscs first appeared in the Lower Cambrian. Mollusc fossil groups include bivalves, gastropods and cephalopods. Molluscs have a muscular 'foot' enabling limited movement, and most have a calcareous shell.

Matherella

Helcionella

Helcionella

Brachiopods

The earliest examples of the brachiopods are bivalved (two-shelled) animals, the shells being held together by muscles attached to the interior of each. The shells themselves are made of layers of chitin and calcium phosphate and are small and conical.

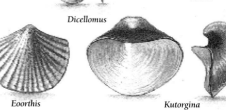

Lingulepis

Paterina

Obolus

Dicellomus

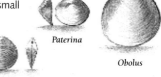

Eoorthis

Kutorgina

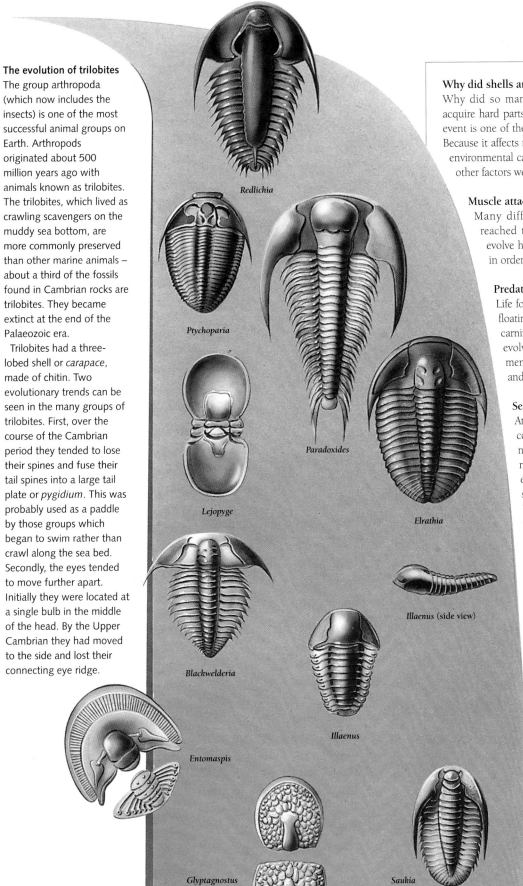

The evolution of trilobites

The group arthropoda (which now includes the insects) is one of the most successful animal groups on Earth. Arthropods originated about 500 million years ago with animals known as trilobites. The trilobites, which lived as crawling scavengers on the muddy sea bottom, are more commonly preserved than other marine animals – about a third of the fossils found in Cambrian rocks are trilobites. They became extinct at the end of the Palaeozoic era.

Trilobites had a three-lobed shell or *carapace*, made of chitin. Two evolutionary trends can be seen in the many groups of trilobites. First, over the course of the Cambrian period they tended to lose their spines and fuse their tail spines into a large tail plate or *pygidium*. This was probably used as a paddle by those groups which began to swim rather than crawl along the sea bed. Secondly, the eyes tended to move further apart. Initially they were located at a single bulb in the middle of the head. By the Upper Cambrian they had moved to the side and lost their connecting eye ridge.

Redlichia

Ptychoparia

Lejopyge

Paradoxides

Elrathia

Blackwelderia

Illaenus (side view)

Illaenus

Entomaspis

Glyptagnostus

Saukia

Why did shells and skeletons evolve?

Why did so many lines of invertebrate animals acquire hard parts at the same time? This dramatic event is one of the hardest for geologists to explain. Because it affects many different types of animal, an environmental cause is likely, though a number of other factors were probably involved.

Muscle attachment

Many different animals simultaneously reached the point where they needed to evolve hard structures to anchor muscles, in order to gain greater mobility.

Predators

Life forms in the Precambrian had been floating plant-eaters. By the Cambrian a carniverous, predatory way of life was evolving. This hastened the development of protective shells, simple jaws and a more active life-style.

Sea level changes

At the end of the Precambrian a vast continent was breaking up to form new oceans, and sea levels were rising as a period of glaciation ended and warm shallow seas spread over continental shelves. These conditions were ideal for marine invertebrate life which evolved and spread rapidly around the continental margins.

Chemistry of the sea

There were striking changes in the composition of seawater between the Precambrian and Cambrian periods. The Precambrian seas contained an excess of lime (calcium carbonate) and this was often precipitated out as lime-rich sediment. At the beginning of the Cambrian the amount of magnesium in the seas decreased rapidly and the dominant mineral changed from aragonite to calcite. At the same time the levels of phosphate in shallow seas increased. All these events are likely to have given opportunities for increase in mineralization and therefore in the formation of shells and skeletons.

The Ordovician World

The Earth 500 to 440 million years ago

The Ordovician period is named after rock formations in the Arenig Mountains of North Wales which were formerly inhabited by a Celtic tribe called the *Ordovices*. The geological time scale for the Ordovician is based on the distinctive changes in graptolite fossils found in the rock strata of the Arenig Mountains. Many of the plants and animal groups that had appeared in the Cambrian era diversified in the warm shallow seas of the Ordovician continental shelves. The earliest land plants may have evolved in the Ordovician – fossilized fragments have been found in rocks in Poland.

Late in the Ordovician period the first of the great Paleozoic mountain-building episodes began. The Taconian Orogeny, named after the Taconic Mountains of New York State, affected south-eastern Canada and the north-eastern United States. The cause was probably the collision of a volcanic island arc with the margin of the continent.

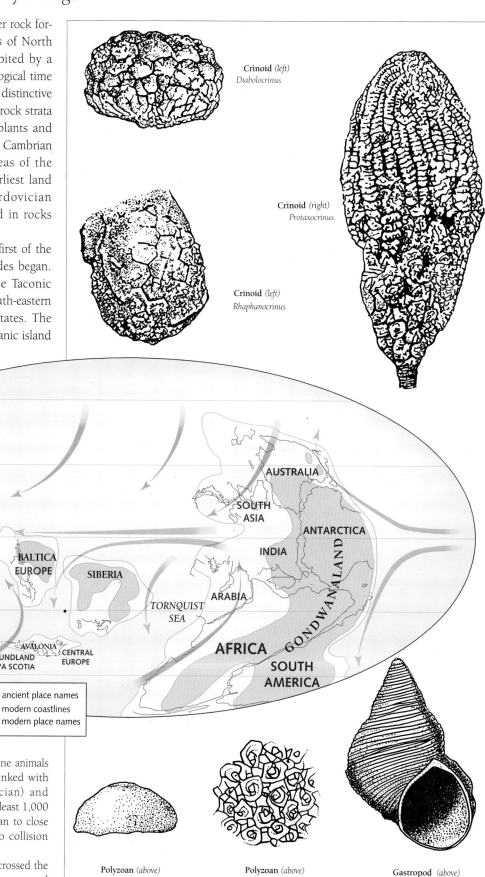

Crinoid (*left*)
Diabolocrinus

Crinoid (*right*)
Protaxocrinus

Crinoid (*left*)
Rhaphanocrinus

THE ORDOVICIAN WORLD

- ancient continents
- ancient continental shelf
- warm ocean currents
- cold ocean currents

LAURENTIA ancient place names
            modern coastlines
SIBERIA   modern place names

AUSTRALIA
SOUTH ASIA
ANTARCTICA
INDIA
ARABIA
AFRICA
SOUTH AMERICA
GONDWANALAND
TORNQUIST SEA
SIBERIA
BALTICA EUROPE
IAPETUS OCEAN
LAURENTIA
NORTH AMERICA
GREENLAND
AVALONIA
NEWFOUNDLAND & NOVA SCOTIA
CENTRAL EUROPE

Polyzoan (*above*)
Prasopora

Polyzoan (*above*)
Prasopora

Gastropod (*above*)
Cyclonema

The Ordovician World

In the Early Ordovician distinct groups of marine animals suggest that Laurentia, Baltica (which was linked with Avalonia over the course of the Ordovician) and Gondwanaland were separated by oceans of at least 1,000 km (600 miles) wide. The Iapetus Ocean began to close later in the Ordovician period, leading up to collision between these continents in the Silurian period.

In the Late Ordovician West Gondwanaland crossed the South Pole. By the end of the period, an ice-cap covered the whole of what is now North Africa.

Major Fossil Groups

The Ordovician period shows the continuing development and diversification of shelled marine invertebrates *(left and below)*. The first vertebrates and the first freshwater and land plants also appeared.

The habitats available for marine plants and animals seemed to increase markedly. Brachiopods, molluscs and graptolites became abundant. Graptolites are found throughout the sediments of the Ordovician and are used as the zone fossil of the period (see *pp. 44–45*). They were small organisms that formed into colonies. They secreted an *exoskeleton* (external skeleton) consisting of one or more *stipes* or branches radiating out from a stem.

Fragmented fossils of the first definite vertebrates – jawless feshwater fish (ostracoderms) have been found in Ordovician rocks in North America. It is known that ostracoderms fed off freshwater plants, so their presence in the Ordovician can be assumed. Broken fossils of the first land plants have been found in Ordovician rocks in Poland and the south-eastern United States. However, the dominant plants of the time continued to be marine reef-building algae.

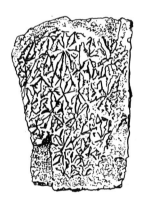

Echinoderm *(above)*
Aulechinus grayae

Echinoderm *(left)*
Heliocrinites

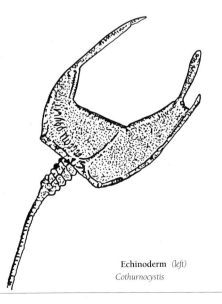

Echinoderm *(left)*
Cothurnocystis

Ordovician Rock Types

Carbonates are abundant on the cratons of Laurentia, Siberia and the India-Australian region of Gondwanaland in the early Ordovician. Sandstones were formed in Baltica and the North African margin of Gondwanaland. In Late Ordovician strata carbonates appear in Baltica, suggesting a move northwards into warm equatorial latitudes. Glacial deposits – tillites and boulder clays – are found in North Africa, giving evidence of Late Ordovician glaciation.

The Silurian World
The Earth 440 to 410 million years ago

The Silurian period was a pivotal time in the Earth's history. At the beginning, marine invertebrate life had become well-established and abundant, while the first vertebrates had appeared in the seas and a few plants had made the transition onto land. By the end of the Silurian, vertebrate fish were widespread and plants and invertebrate animals had establised a definite presence on land.

Meanwhile the major event of the Silurian period in the northern continents was the culmination of the Caledonian Orogeny (*see page 60*). This complex series of events resulted from the closing of the ocean basin that lay between Laurentia (the core of what is now North America), Baltica (Scandinavia and northern Britain), and Avalonia (southern Britain, Nova Scotia and Newfoundland). The Caledonian mountain chains of north-east North America and north-west Europe are the much-eroded remnants of this tectonic event.

Geologists had relatively easy access to these well exposed mountains across northern Europe and the north-east United States, and they were among the earliest rock formations to be studied. Once the implications of continental drift were accepted, it became apparent that the mountains on either side of the North Atlantic were the result of the same geological event.

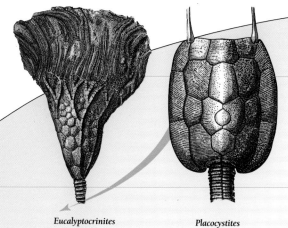

Eucalyptocrinites

Placocystites

Crinoids
Crinoids (*left*) emerged in the Cambrian, and became widespread in the Silurian. They were stemmed echinoderms, with the organism living inside the bowl. Tentacles were used for feeding. In some groups the stems broke off and the crinoid became free-floating.

SIBERIA

LAURENTIA

NORTH AMERICA

AVALONIA

FLORIDA

Pharyngolepis

Jamoytius

Dartmuthia

Silurian Fish
Jawless fish (*left*) fed on microscopic organisms. Unlike their modern equivalents, some Silurian jawless fish developed head armour - like this 10 cm (4 in)-long *Dartmuthia*, from the late Silurian period.

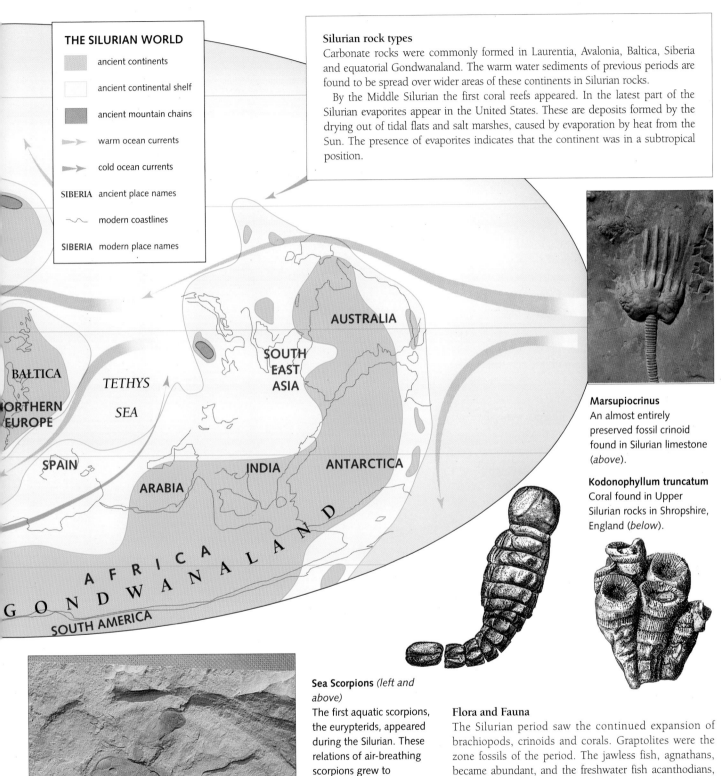

THE SILURIAN WORLD

ancient continents

ancient continental shelf

ancient mountain chains

warm ocean currents

cold ocean currents

SIBERIA   ancient place names

modern coastlines

SIBERIA   modern place names

Silurian rock types

Carbonate rocks were commonly formed in Laurentia, Avalonia, Baltica, Siberia and equatorial Gondwanaland. The warm water sediments of previous periods are found to be spread over wider areas of these continents in Silurian rocks.

By the Middle Silurian the first coral reefs appeared. In the latest part of the Silurian evaporites appear in the United States. These are deposits formed by the drying out of tidal flats and salt marshes, caused by evaporation by heat from the Sun. The presence of evaporites indicates that the continent was in a subtropical position.

Marsupiocrinus
An almost entirely preserved fossil crinoid found in Silurian limestone (*above*).

Kodonophyllum truncatum
Coral found in Upper Silurian rocks in Shropshire, England (*below*).

Sea Scorpions (*left and above*)
The first aquatic scorpions, the eurypterids, appeared during the Silurian. These relations of air-breathing scorpions grew to monstrous sizes – up to 3 metres'(10 ft). Eurypterids had two pairs of eyes, four pairs of walking legs, one pair of oar-like legs and one pair of pincers. They died out at the end of the Permian period.

Flora and Fauna

The Silurian period saw the continued expansion of brachiopods, crinoids and corals. Graptolites were the zone fossils of the period. The jawless fish, agnathans, became abundant, and the freshwater fish acanthodians, the first fish with jaws, appeared. The first arachnids and centipedes appeared in the Upper Silurian.

Silurian rocks contain the earliest fossils of vascular land plants – the 5 cm (2 in) – tall Cooksonia. Fossils of two important groups of land plants (lycopsids and psilopsids) have been found in Upper Silurian rocks in Australia. These were the precursors of the great forests of the Carboniferous.

The Caledonian Mountains
Europe and North America in collision

The mountains of northern Britain, Norway and Spitzbergen are part of a continuous chain with strong geological affinities to the mountains of south-eastern Canada and the north-eastern United States. These mountains were all formed during the early Paleozoic by a series of events known as the Caledonian Orogeny. Work on the Caledonian Mountains in the 18th century inspired scientists to propose ideas that were the foundations of modern geology.

Though the Caledonian Orogeny is a particular geological event it took place over an extremely long time-scale, even by geological standards. Around 425 million years ago the continental blocks were mainly in the southern hemisphere. At this time the separation of Laurentia (comprising present-day North America) from Africa had reached its widest point, and the continents were about to begin the long process of joining together.

The small continents of Baltica (comprising Scandinavia and the southern British Isles) and Avalonia (comprising Newfoundland, Nova Scotia and northern Britain), were being pushed against the north-eastern edge of Laurentia by the spreading Iapetus Ocean located to the south and east. This collision caused disturbance and mountain-building along the nothern edge of Baltica, forming the Caledonian mountains of Norway, Scotland, and eastern Canada.

By about 400 million years ago the Iapetus Ocean had begun to close. But the rotation of the continents opened up a large ocean to the north-east – the Tethys Sea. Gondwanaland and Africa were now moving in a clockwise rotation as well as moving west. This caused the focus of the collision with Laurentia to spread southwards from the Canada/Britain boundary to Avalonia and what is now the north-east United States. Florida, still attached to Gondwanaland, was swinging around towards the south-east margin of the North American continent. Around 360 million years ago, Gondwanaland collided with North America along the whole of its eastern margin. This caused the first uplift of the Appalachian region, a process that was to continue for the next 100 million years.

The southern continents (*left*) were joined together in Gondwanaland which was centred on the South Pole.

The Iapetus Ocean (*below*) had reached its maximum extent and was about to begin closing.

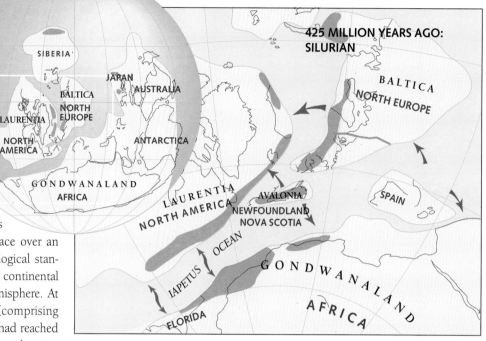

THE WORLD
425 MILLION YEARS
AGO – SILURIAN

425 MILLION YEARS AGO:
SILURIAN

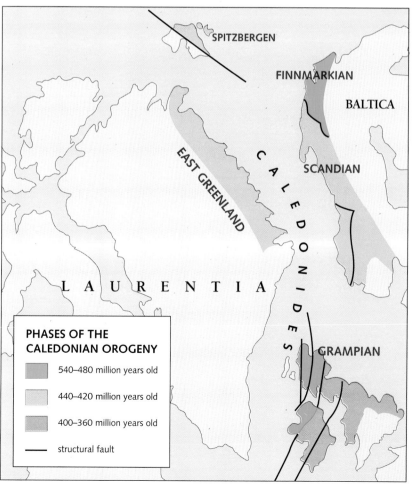

PHASES OF THE
CALEDONIAN OROGENY

540–480 million years old

440–420 million years old

400–360 million years old

structural fault

Gondwanaland (*below*),
moving clockwise, has
collided with Laurentia,
opening up the Tethys Sea.

All the continents have
moved north, and
Gondwana and Laurentia
are in collision along the

length of North America,
causing massive disturbance
and mountain-building.

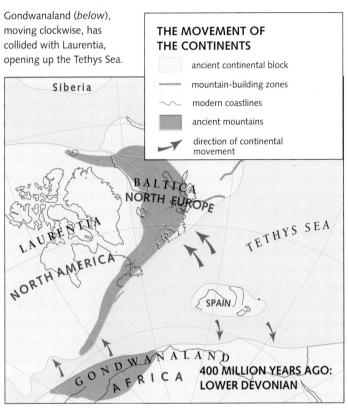

**THE MOVEMENT OF
THE CONTINENTS**

- ancient continental block
- mountain-building zones
- modern coastlines
- ancient mountains
- direction of continental movement

**400 MILLION YEARS AGO:
LOWER DEVONIAN**

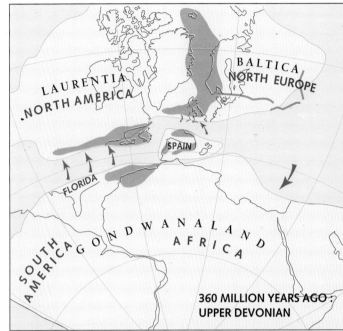

**360 MILLION YEARS AGO:
UPPER DEVONIAN**

Caledonia

The great mountain-building and volcanic episodes that occurred early in the Paleozoic era around the North Atlantic are all related to the same tectonic events.

Volcanic episodes occurred in the Caledonian belt (*map left*) as early as the Late Cambrian period (540 million years ago). These episodes intensified during the Ordovician period, and peaked during the Caledonian Orogeny in the Silurian (400 million years ago).

Peaks of activity occurred in different places at different times. The oldest Caledonian rocks are found in northern Norway in the Finnmarkian formation, perhaps caused by the collision of Baltica with an island arc. Further south the Grampian Mountains were formed by collision between the southern British Isles and north-west Scotland, then attached to Laurentia off the east coast of Greenland.

An interval of perhaps 50 million years passed before the main activity in the northern part of the Caledonian zone began in earnest around 430 million years ago. The Scandian and East Greenland zones overlap in age, and were presumably formed in the same episode, as was the eastern part of the island of Spitzbergen.

Finally, in this complex set of movements, the disturbance zone moved south again. The uplands of southern Scotland, northern England, Wales and Ireland were formed 400 to 360 million years ago in the final phase of the Caledonian Orogeny.

Caledonian mountains stretch further south to eastern Canada. But they are overlain by later events such as the Appalachian Orogeny. The combined episodes created an almost continuous chain of mountains from Newfoundland to Tennessee.

Caledonian Mountains
The mountain landscapes of Glencoe in Scotland (*above*) and northern Norway (*left*) were formed around the same time. The similarities in age and composition of the rocks gives the mountains a similar appearance, matching those across the Atlantic.

The Devonian World
The Earth 410 to 350 million years ago

New mountain chains had been formed on the margins of the northern continents in the previous Silurian period. They were rapidly eroded during the Devonian, and huge quantities of sandstone were deposited in great thicknesses over a large area. This Old Red Sandstone is the characteristic rock of this period on the northern continents.

The Devonian period featured a major diversification of fishes, and a growth in abundance of these marine vertebrates. Some developed air-breathing lungs, and other groups evolved legs. The lungfish are the precursors of amphibians – significant steps in the vertebrate move onto the land. Land plants were established at the beginning of the Devonian, and by the end of the period there were dense coastal forests containing a diversity of groups.

In the Northern Hemisphere the linking of Laurentia, Avalonia and Baltica explains the presence of Old Red Sandstone in present-day North America, Greenland and Europe. Warm water limestones appear in Devonian rocks in Morocco, Brittany and south-west England.

THE LOWER DEVONIAN WORLD

SIBERIA

NORTHERN IAPETUS OCEAN

GREENLAND

LAURENTIA/BALTICA

NORTH AMERICA

EUROPE

KAZAKHSTAN

SOUTH ASIA

AUSTRALIA

SOUTH AMERICA

AFRICA

ARABIA

INDIA

ANTARCTICA

G O N D W A N A L A N D

The first sharks, such as the 75 cm (30 in) long *Xenacanthus*, and armoured fish such as *Dunkleosteus*, swam in Devonian seas.

The Devonian World
Throughout the Devonian era western Gondwana and Laurentia continued to move northward. Closure of the Northern Iapetus Ocean led to mountain building in northern Norway and eastern Greenland

The Age of Fishes

Freshwater and marine varieties of fishes proliferated in the Devonian. Varieties included several kinds of armoured fish (placoderms), jawless fish and the first bony fish. Sharks first appeared in the Middle Devonian, making them one of the longest lived groups of vertebrate animals still in existence. Fish with true jaws (acanthodians) appeared in the Upper Silurian and/or Lower Devonian.

Other life forms making their first appearance were primitive air-breathing fish. These are the link between fish and the first amphibians, which had emerged by the end of the Devonian. In the Late Devonian some amphibians, which had four legs bearing toes, became the first vertebrates to walk on the land (see *pp. 64–65*).

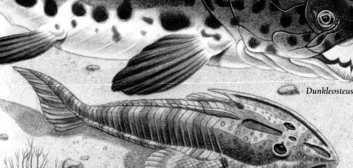

Xenacanthus

Dunkleosteus

Agnathans

Thamnopora Coral

The Devonian period saw the first great reefs built by corals and other marine animals.

Flora and fauna

Brachiopods, corals, stromatoporoids, crinoids, trilobites and gastropods continued to be the dominant marine invertebrate groups. Fossils of the giant sea scorpions (eurypterids) that emerged in the Silurian are also found in Old Red Sandstone. The first spiders and insects appeared in the Devonian. A great spread of insects had occurred by the mid-Devonian, but fossil insect finds are very rare.

In the mid-Devonian there was an explosion of land plant life, in particular a proliferation of fern-like plants. Early gymnosperms, seed-bearing plants, (for example, *Callixylon*) occur in the Devonian and lycopsids were abundant by the Late Devonian.

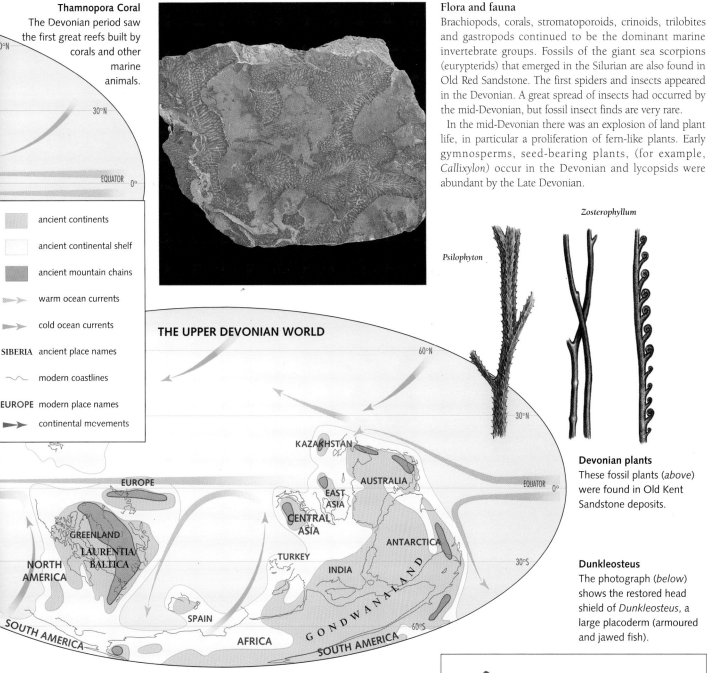

ancient continents

ancient continental shelf

ancient mountain chains

warm ocean currents

cold ocean currents

SIBERIA ancient place names

modern coastlines

EUROPE modern place names

continental movements

THE UPPER DEVONIAN WORLD

Psilophyton

Zosterophyllum

Devonian plants
These fossil plants (*above*) were found in Old Kent Sandstone deposits.

Dunkleosteus
The photograph (*below*) shows the restored head shield of *Dunkleosteus,* a large placoderm (armoured and jawed fish).

Devonian rock types

As well as the usual coastal and marine sediments, there were also widespread continental and desert-type deposits in the Devonian period. Devonian rocks are found in all continents and are valuable sources of iron ore, tin, zinc, and copper, and of oil and evaporites.

In north-west Europe and eastern North America great masses of sand and mud accumulated in structural basins between the ranges of the Caledonian Mountains. In Europe these non-marine, mainly river deposits are known as Old Red Sandstone. They match exactly the red beds of the Catskill Mountains of New York State. The sediments actually consist of red, green and grey sandstones and grey shales in a poorly sorted mixture.

Life onto Land

Plants and animals move from sea to land

For the first 4,000 million years of the Earth's history life existed only in the seas. Then, about 420 million years ago, the first plants began to colonize the land. They were followed 40 million years later by the first land vertebrates – amphibians with the ability to live on land, but still tied to water. The first true land-dwelling vertebrates were the reptiles, which emerged during the Carboniferous period about 300 million years ago. Invertebrates probably colonized the land before the vertebrates, but fossil finds have proved elusive.

But why did life come onto land in the first place? The answer may be sunlight. Plants need sunlight to manufacture their food through photosynthesis, and plants on land can get more sunlight than those in water. Plants had to undergo a massive adaptation to survive in the new environment. In water dispersal, plants use water currents for reproduction. The first land plants to develop a different reproductive system were the gymnosperms. Each pollen grain of a gynmosperm carries with it a watery environment in the form of a tube through which sperm swim to the egg. The fertilized egg becomes a seed – the most successful method of reproduction for land plants. Once land plants became established, animals followed this untapped food source. Both plants and animals then evolved to take advantage of the opportunities the land offered.

LUNGFISH IN THE UPPER DEVONIAN WORLD

- ancient continents
- ancient continental shelf
- ancient mountain chains
- • Early Devonian lungfish sites
- ○ late Devonian lungfish sites

Ichthyostega
Marks the transition from fish to amphibian (*below*). Fossils occur in Upper Devonian rocks.

Life cycle of seedless vascular plants (*left*)
Vascular plants colonized the land, but were still tied to water by their method of reproduction, and spores are transported from male to female in water.

spores
adult plant
Gametophyte
young plant
fertilization in water
sperm

Transitional forms

It took animals over 100 million years to evolve from sea to land. The first stage was the development of lungfish (*see map*). Amphibian forms such as *Ichthyostega* (*above*) followed. These were able to survive out of water for long periods, enabling them to feed on the newly developed land plants. But they were tied to water for reproduction. The next stage was the emergence of true reptiles. *Seymouria* (*top of page*) is a transitional form between amphibians and reptiles from 280 million years ago. Reptiles lay eggs which can be fertilized and incubated on land, freeing them forever from dependence on water.

Myxinoidei
Petromyzontoidei
Acanthodii
Ostoestraci
Anaspida
AGNATHANS
GNATHOSTOMES
Heterostraci
Placodermi

Lungfish sites

In the Devonian period some fish groups began to develop lungs as an alternative method of absorbing oxygen. The map shows fossils finds of lungfish, plotted on a reconstruction of the Devonian world.

Seymouria

An amphibian from the Permian period, *Seymouria* (*above*) showed many reptile like characteristics.

Plants transform the land

In the early Devonian (*above right*) only a few plant species, such as ground-creepers, could survive on land, and they were tied to wet habitats. By the late Devonian (*right*) lowlands were covered by large forests, as plants developed new methods of reproduction using seeds.

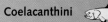

Rhipidistia

Amphibia

Coelacanthini

Dipnoi

Actinopterygii

Elasmobranchii

Holocephali

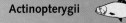

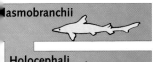

From fish to amphibians

The earliest fish, from the Ordovician period, had external skeletons formed as armour around soft bodies, with perhaps an internal skeleton of cartilage. These agnathans were jawless sediment-feeders. The first jawed fish (placoderms) arrived in the Silurian. They died out at the end of the Devonian period, but the development of jaws opened the way for the most successful fish group, the Osteichthyes. These had bony skeletons, and their body armour was reduced to thin scales. Some fish in this group developed nostrils and lungs, others had muscular fins which allowed them to 'walk' on the seafloor.

Gilboa Forest

The earliest fossil of a vascular plant dates from the Late Silurian in Wales. Once established on land, plants evolved very rapidly in the Devonian period. By the Late Devonian there were recognizable forests – the Gilboa Forest was discovered in the Upper Devonian strata of the Catskill Mountains in New York State. Lying along a low river valley, this forest contained trees over 10 metres (30 ft) high. These trees belonged to new plant groups – the lycopsids and sphenopsids.

The true ferns (Filinacae), which appeared in the Late Devonian, were another important early land plant. These ferns grew up to 20 metres (65 ft) high. The true ferns reproduce by spores, whereas the similar-looking seed ferns (pteridosperms) produce seeds. The seed ferns, which are common in the Gilboa Forest, are the probable ancestors of the gymnosperms and the later flowering plants.

Part Four
The Late Paleozoic Era

The Lower Carboniferous World
The Earth 350 to 315 million years ago

The Carboniferous period is named after the carbon-rich coal measures that are characteristic of the Upper Carboniferous in Europe and North America. In the northern continents there is a distinct change of characteristics half way through the period, which is therefore divided by geologists into the Upper and Lower Carboniferous – in North America these are known as the Mississippian and Pennsylvanian divisions.

The Lower Carboniferous world was dominated by three continental blocks. Lying well within the tropics Laurentia, made up of North America, north-west Europe, Greenland and Russia west of the Urals, remained as an equatorial continent throughout the period. Further north Angaraland, comprising Asiatic Russia, Siberia and parts of China lay partly in the temperate zone and stretched away towards the Arctic. The present day southern continents were joined together in the huge continent of Gondwanaland. Much smaller fragments of continents lay between these major blocks. The most geologically active region was the boundary between Laurentia and Gondwanaland. As Laurentia turned anti-clockwise it collided with Gondwanaland along the Appalachian mountain belt.

On the northern continents the Lower Carboniferous was a time of transition from the mainly marine conditions of the early Paleozoic, to the swamps of the Upper Carboniferous and the dry land conditions of the Permian period.

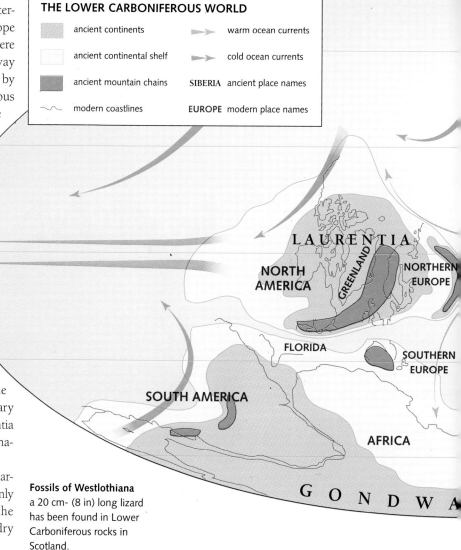

THE LOWER CARBONIFEROUS WORLD

- ancient continents
- ancient continental shelf
- ancient mountain chains
- modern coastlines
- warm ocean currents
- cold ocean currents
- SIBERIA ancient place names
- EUROPE modern place names

LAURENTIA

NORTH AMERICA GREENLAND NORTHERN EUROPE

FLORIDA SOUTHERN EUROPE

SOUTH AMERICA

AFRICA

GONDWA

Fossils of Westlothiana
a 20 cm- (8 in) long lizard has been found in Lower Carboniferous rocks in Scotland.

Meganeura

Carboniferous insects
Insects and arachnids (spiders) emerged in Silurian and Devonian times. By the Carboniferous they were well developed and diversified. *Plesiosora* was an early arachnid. *Aphthoroblattina*, a large cockroach, and *Meganeura*, a dragonfly with a wingspan of 50 cm (18 inches).

Apthoroblattina

Plesiosora

Fauna of the Lower Carboniferous
The first true reptiles appeared in the Carboniferous. *Westlothiana* (*above*) is the oldest known reptile.

Marine animals such as brachiopods and ammonoids flourished throughout the Carboniferous and into the Permian. Corals declined in importance, and massive reefs were built up from stemmed creatures known as crinoids. Crinoids and blastoids are part of the echinoderm family, which includes starfish. They diversified greatly in the Lower Carboniferous.

The goniatites group of ammonoids appeared, and were to last until the mass extinction at the end of the Permian period. The microscopic Foraminifera diversified rapidly, and are used by geologists as indicators. They are invaluable because of their rapid evolutionary patterns – some were free-floating, and therefore dispersed rapidly, and are easily preserved as fossils.

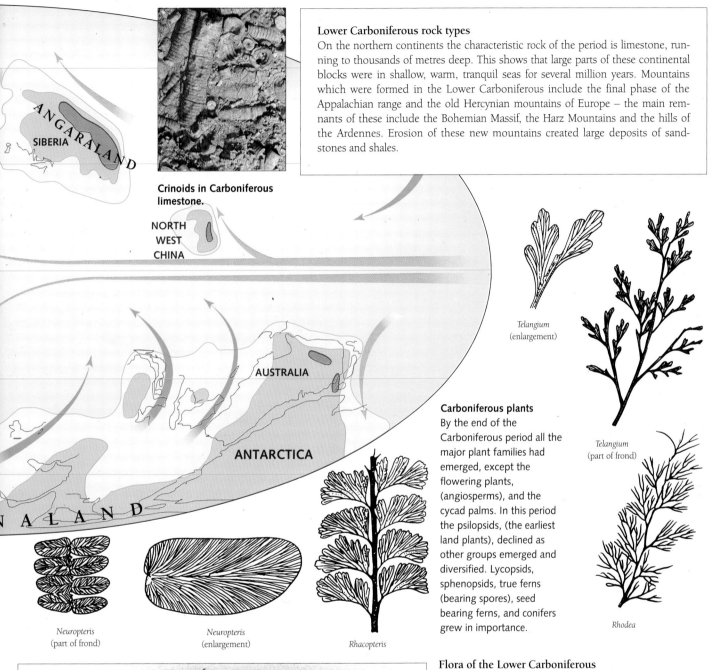

Lower Carboniferous rock types

On the northern continents the characteristic rock of the period is limestone, running to thousands of metres deep. This shows that large parts of these continental blocks were in shallow, warm, tranquil seas for several million years. Mountains which were formed in the Lower Carboniferous include the final phase of the Appalachian range and the old Hercynian mountains of Europe – the main remnants of these include the Bohemian Massif, the Harz Mountains and the hills of the Ardennes. Erosion of these new mountains created large deposits of sandstones and shales.

Crinoids in Carboniferous limestone.

Telangium (enlargement)

Telangium (part of frond)

Carboniferous plants

By the end of the Carboniferous period all the major plant families had emerged, except the flowering plants, (angiosperms), and the cycad palms. In this period the psilopsids, (the earliest land plants), declined as other groups emerged and diversified. Lycopsids, sphenopsids, true ferns (bearing spores), seed bearing ferns, and conifers grew in importance.

Rhodea

Neuropteris (part of frond)

Neuropteris (enlargement)

Rhacopteris

Stethacamthus, a Carboniferous shark, was 70 cm (30 in) long. It had a toothed fin on its back, possibly used in display.

Flora of the Lower Carboniferous

There were three distinct groups of flora in the Carboniferous, each associated with one of the major continental blocks. The tropical and subtropical zone was dominated by lycopsid plants until the mid to late Carboniferous, when ferns began to take over. This flora is found in Carboniferous rocks in North America, Europe, North Africa and China.

A zone of temperate flora is found in Carboniferous rocks in eastern Siberia and includes early conifers with annual growth rings (indicating a seasonal climate), as well as seed ferns and lycopsids. The southern cool temperate flora is named after the *Glossopteris* seed fern, which is its dominant genus. The *Glossopteris* flora is associated with glacial deposits and contains tree trunks with prominent seasonal growth rings. It also produced vast coal deposits.

Limestone
Chemical deposition of carbonates

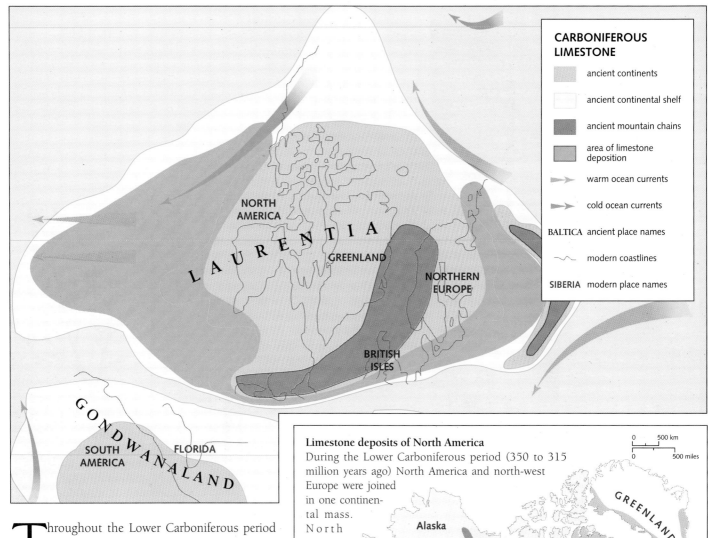

Throughout the Lower Carboniferous period the northern continents were partially submerged in warm, shallow tropical seas. We know this because of the thick deposits of limestone found all over North America and northern Europe. The huge beds of limestone are often interleaved with thinner bands of sandstone and shale which show that the shallow seas periodically receded to create delta and coast conditions. In some places the build-up of limestone reached great depths, demonstrating that the conditions for deposition were remarkably long lasting. Gradual subsidence of the shallow sea floor was needed to allow such thick sediments to build up.

Limestones, or carbonates, are formed by chemical depostion. When a body of water contains more carbonates than it can dissolve, they are precipitated out, either as shells made by marine animals, or as minute mineral particles. These then settle on the sea floor and are compressed into rock in the same way as sandstones, shales and mudstones.

Limestone deposits of North America

During the Lower Carboniferous period (350 to 315 million years ago) North America and north-west Europe were joined in one continental mass. North America was rotated clockwise relative to its present position, and the continent straddled the equator. Sea levels fluctuated throughout the period, though large parts of the continent were flooded by warm shallow seas for much of the time. Only high land in eastern Canada, Greenland, Scandinavia and Scotland remained above sea level for the whole period. The map (*top*) shows the average coastline. Limestone deposition occurred in shelf areas.

The present-day map of limestone deposits from the time (*right*), shows how these reflect the geography at the time of deposition.

Formation of Limestone

The main way in which carbonates are taken out of ocean water is by marine organisms forming shells. When marine organisms die their shells sink to the bottom of the oceans and are buried and compressed to form limestone. Limestones, in general, are made from the hard body parts of countless millions of marine organisms.

Shelly limestone (*far left*)
In this limestone, fragments of large shells and skeletons from crinoids, corals and trilobites can be seen by the naked eye. The large shells are bound together by microscopic ooliths or by carbonate mud.

Oolitic limestone (*left*)
Ooliths are minute fragments of shell or sand which have been rolled about on the sea bed, and become coated in calcium carbonate. The resulting spheres are held together by calcite.

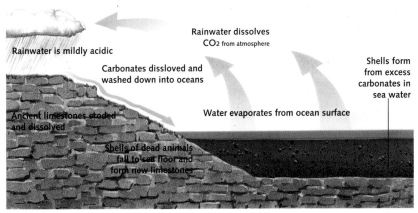

The Carbon Cycle

Rainwater contains some dissolved carbon dioxide and acts as a weak acid. When rocks are eroded some minerals – principally carbonates – are dissolved by the weak acid. The solution is transported by rivers down to the oceans. Water continually evaporates from the oceans' surface and is replenished by fresh river water bringing in more minerals. These minerals are kept in chemical balance by chemical precipitation – as carbonates are washed into one place, they must be precipitated out in another. The warm waters of the tropics are heavily saturated in carbonates, and it is in these regions that most carbonate deposition takes place, and where most limestones are formed.

Limestone cliff (*above*)
The region of the English Pennines was subject to lengthy submergence by shallow seas during the Lower Carboniferous period. The large quantities of limestone formed then were later uplifted to form thick outcrops like this cliff-face at Malham in Yorkshire.

Limestone pavement (*left*)
Glaciers and ice-sheets have stripped the overlying soil and sediment off this limestone rock, again at Malham. Further erosion has left a characteristic pavement effect.

The Upper Carboniferous World

The Earth 315 to 290 million years ago

In the northern continents the beginning of the Upper Carboniferous, or Pennsylvanian, period saw a change from a shallow marine environment to huge coastal swamps and deltas. In the south conditions became colder and an ice sheet began to extend across the polar portion of Gondwananland.

New plants devloped to take advantage of the warm wet conditions in Laurentia and northern Europe. These plants then contributed to the formation of swamp forests, which provided yet more diverse conditions in which plants and animals thrived. The swamps were particularly suitable for the diversification of insects and reptiles. A different type of flora spread over the south in cool temperate conditions. Both flora formed large coal deposits.

Another step in the formation of the supercontinent of Pangea was made with the joining of Laurentia and Gondwanaland. These massive continental blocks were locked together for the remainder of the Paleozoic era. Gondwanaland moved over the South Pole bringing lower temperatures, and eventually ice-sheets, to the southern continents. For part of the period an ice-sheet extended from Argentina across south-central Africa to Madagascar, southern Arabia, India, Antarctica and Australia.

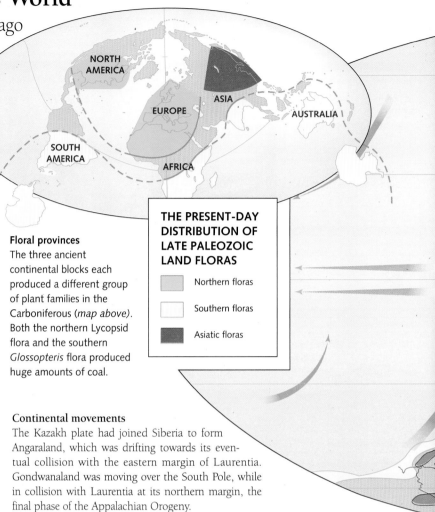

THE PRESENT-DAY DISTRIBUTION OF LATE PALEOZOIC LAND FLORAS

- Northern floras
- Southern floras
- Asiatic floras

Floral provinces
The three ancient continental blocks each produced a different group of plant families in the Carboniferous (*map above*). Both the northern Lycopsid flora and the southern *Glossopteris* flora produced huge amounts of coal.

Continental movements
The Kazakh plate had joined Siberia to form Angaraland, which was drifting towards its eventual collision with the eastern margin of Laurentia. Gondwanaland was moving over the South Pole, while in collision with Laurentia at its northern margin, the final phase of the Appalachian Orogeny.

Marine vertibrates

Megalichthys, dorsal view

Megalichthys, ventral view

Megalichthys, restored skull

Megalocephalus

Upper Carboniferous marine vertebrates
Fish had become well established in the Devonian period, and continued to thrive in the Carboniferous. Jawed, bony fish with scales had largely replaced jawless armoured fish. This *Megalichthys* (*left*) has been preserved in Upper Carboniferous rock near Wakefield, Yorkshire.

Although reptiles arose in the Carboniferous, amphibians continued to be common, particularly in swamp areas. This beautifully preserved *Megalocephalus* (*bottom*) was fossilized in Dawley, Shropshire.

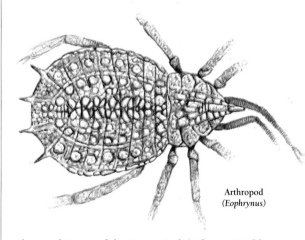

Arthropod (*Eophrynus*)

Flora and Fauna of the Upper Carboniferous World
The major change in life forms in the Carboniferous is related to the development of coastal swamps. This involved the evolution of new types of plant, such as the *Lepidodendron* and *Sigillaria* groups. As these plant groups became established new animals evolved to take advantage of the changes in habitat. The swamps were an ideal environment for the development of insects and these diversified greatly during the Carboniferous. Flying insects developed, including dragonflies with wings spans of over 15 cm (6 in) and 10 cm (4 in)-long cockroaches.

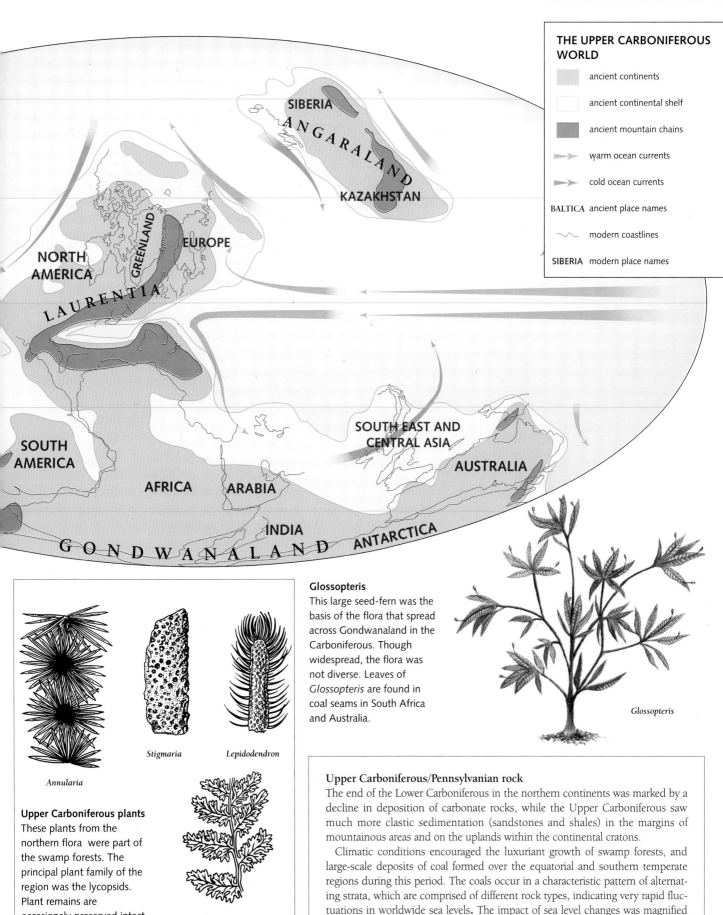

THE UPPER CARBONIFEROUS WORLD

- ancient continents
- ancient continental shelf
- ancient mountain chains
- ⟶ warm ocean currents
- ⟶ cold ocean currents
- BALTICA ancient place names
- ∿ modern coastlines
- SIBERIA modern place names

SIBERIA
ANGARALAND
KAZAKHSTAN

NORTH AMERICA
GREENLAND
EUROPE
LAURENTIA

SOUTH AMERICA

AFRICA   ARABIA

SOUTH EAST AND CENTRAL ASIA

AUSTRALIA

INDIA   ANTARCTICA
G O N D W A N A L A N D

Annularia

Stigmaria

Lepidodendron

Upper Carboniferous plants
These plants from the northern flora were part of the swamp forests. The principal plant family of the region was the lycopsids. Plant remains are occasionaly preserved intact in coal measures.

Sphenopteris

Glossopteris
This large seed-fern was the basis of the flora that spread across Gondwanaland in the Carboniferous. Though widespread, the flora was not diverse. Leaves of *Glossopteris* are found in coal seams in South Africa and Australia.

Glossopteris

Upper Carboniferous/Pennsylvanian rock
The end of the Lower Carboniferous in the northern continents was marked by a decline in deposition of carbonate rocks, while the Upper Carboniferous saw much more clastic sedimentation (sandstones and shales) in the margins of mountainous areas and on the uplands within the continental cratons.

Climatic conditions encouraged the luxuriant growth of swamp forests, and large-scale deposits of coal formed over the equatorial and southern temperate regions during this period. The coals occur in a characteristic pattern of alternating strata, which are comprised of different rock types, indicating very rapid fluctuations in worldwide sea levels. The impact of sea level changes was magnified by the flatness of the continents.

The Great Swamp Forests
Coal from vegetation in Pennsylvania and Northern Europe

The swamp forests of the Upper Carboniferous, or Pennsylvanian, period were one of the great wonders of the Earth's history. As the continent of Laurentia drifted into the tropics, plants emerged to take advantage of the shallow coastal lagoons. They created forests that stretched from the eastern edge of European Russia, through Poland, northern Germany, the British Isles and across eastern North America. The nearest equivalent today are the Florida Everglades, but the Carboniferous forests were on a continental scale.

The swamp conditions which fostered the growth of the forests also made possible the formation of coal. The waters of the swamp were highly acidic – this prevented dead and fallen trees from rotting. The carbon content of the vegetation was preserved and gradually pressured into seams of coal, instead of being oxidized and lost as carbon dioxide gas. The resulting coal-fields were the cradle of the industrial revolution.

A different type of coal-forming forest was developing at the same time in the southern continents. This contained entirely different species of vegetation in a cool temperate climate. The trees of the southern flora contain annual growth rings, showing large seasonal variations, unlike the tropical forests of the north.

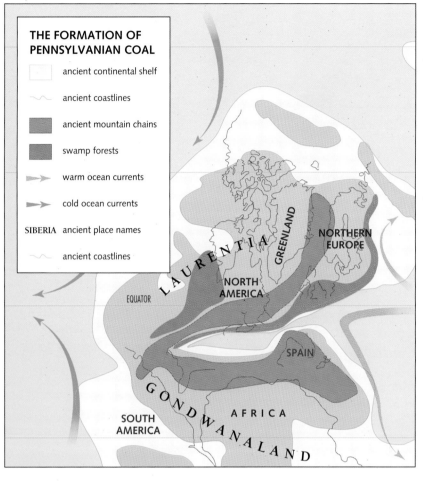

THE FORMATION OF PENNSYLVANIAN COAL

- ancient continental shelf
- ancient coastlines
- ancient mountain chains
- swamp forests
- warm ocean currents
- cold ocean currents
- SIBERIA ancient place names
- ancient coastlines

The formation of coal

In the Upper Carboniferous swamps newly evolved plants like *Lepidodenron* and *Sigillaria* thrived in warm wet conditions. As they died they crashed into the stagnant acidic water of the swamp, which kept them from rotting. Instead, fibres from the roots and stems matted together to form peat. The water and any remaining oxygen was squeezed out of the peat as more peat and then more sediment was piled on top. This leaves hard layers or seams of coal, which are rich in carbon. It takes about 6 metres (20 ft) of peat to make a 1 metre (3 ft)-thick coal seam. Coal seams often have a layer of seat earth beneath them, representing the swamp floor. It contains roots of swamp plants.

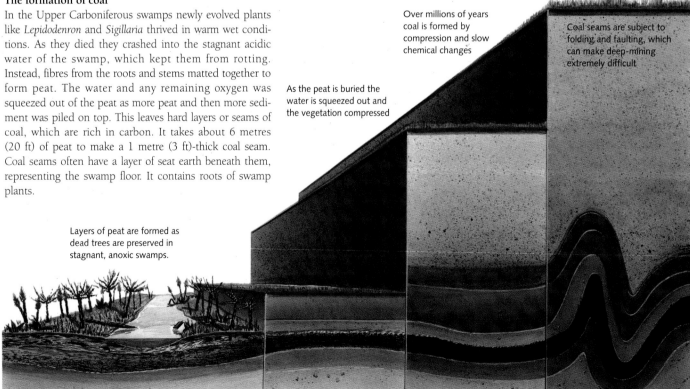

Layers of peat are formed as dead trees are preserved in stagnant, anoxic swamps.

As the peat is buried the water is squeezed out and the vegetation compressed

Over millions of years coal is formed by compression and slow chemical changes

Coal seams are subject to folding and faulting, which can make deep-mining extremely difficult

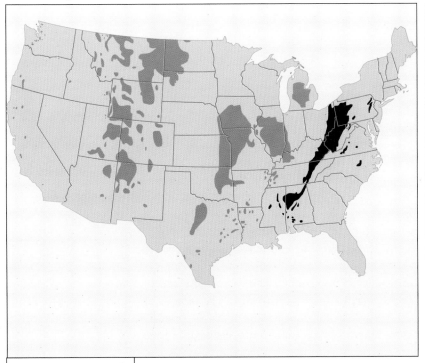

THE COALFIELDS OF NORTH AMERICA

 Pennsylvanian coalfield

 Post-Carboniferous coalfield

Coal in North America
(*above*) The oldest coal anywhere in the world is from the Carboniferous period, as before then plants suitable for conversion into coal did not exist. In the United States the Carboniferous coalfields lie in a band running from Pennsylvania and Ohio south to Tennessee and north Alabama.

In contrast, the coal in the western United States is from the Permian period, and younger. It is poorer quality and lies nearer the surface. It is often mined by open-cast methods (*above*).

A Carboniferous Forest
Low-lying water-logged land, and warm and moist air created perfect conditions for the proliferation of plants on the continental margins of Laurentia and northern Europe.

Huge lycopsid trees dominated the tropical swamps. The two principal species were the tall, branching *Lepidodendron* and the short, broad *Sigillaria*. Insects thrived in the damp warm conditions; dragonflies grew to wingspans of 15 cm (6 in).

The Appalachians
The shaping of North America

The topography of the eastern United States is dominated by the Appalachian Mountain range. It stretches from Newfoundland to Alabama, and the effects of the disturbances that built the Appalachians are seen as far west as the Marathon Mountains of Texas. The formation of the Appalachians resulted from the collision of the two major continental blocks of the time – Laurentia and Gondwanaland.

By the beginning of the Carboniferous period 350 million years ago northern Europe had become joined to Laurentia. Africa and South America, part of the great southern continent of Gondwanaland, had pushed into the south-east margin of Laurentia, completing the first phase of the uplift of the Appalachian region. The Tethys Sea had been created by the conjunction of Gondwanaland with Baltica and northern Europe. The Tethys was to endure for many millions of years, dividing Europe in two. The continent of Siberia was located off to the north-east of Baltica.

As Gondwanaland pushed northwards subduction zones formed around the Tethys Sea. New continental crust was formed in the Tethys – portions of this have endured to the present as the massifs of Central Europe, Bohemia and Pannonia. Further west Gondwana was again in collision with Laurentia 300 million years ago. This renewed disturbance pushed huge quantities of sediment and volcanic material, which were deposited in the Caledonian era, up to 250 kilometres (150 miles)

westwards. This overthrusting, which took place on a continental scale, was responsible for the formation of the central Appalachian Mountain belt. The micro-continent of Florida, which had become attached to Gondwanaland, was in collision with Laurentia at the southern end of the Appalachian belt.

At the climax of the Hercynian Orogeny, of which the Appalachian Orogeny was a part, 250 million years ago, there were disturbances along a whole series of continental boundaries. As Siberia (Angaraland) approached Baltica volcanic acivity began along its south-western edge. The margins of the Tethys Sea continued to be active subduction zones. There was a renewal of the disturbance zone lying between Northern Europe and Greenland as these two continents moved closer together. To the south-west the Appalachian mountain-building entered its final phase. Florida collided with Laurentia as collision, volcanic activity and overthrusting continued along the whole of the southeastern edge of Laurentia.

The Appalachians are a comparatively old mountain chain. In contrast to younger formations of the west coast of America they are presently – and for the foreseeable future – geologically inactive.

THE LOWER CARBONIFEROUS WORLD

The Appalachian Orogeny
The first uplift of the Appalachians took place 350 million years ago, when Africa and South America, part of Gondwanaland, pushed into the south-east margin of Laurentia. Gondwanaland, moving north, caused massive overthrusting, which formed the central Appalachians. When Florida collided with Laurentia 250 million years ago, the Appalachian orogeny was complete.

	ancient continents
	ancient continental shelf
	ancient mountain chains
➤	continental movement

350 MILLION YEARS AGO

SIBERIA

LAURENTIA

NORTH AMERICA

CENTRAL EUROPE

BOHEMIA

TETHYS SEA

SOUTH AMERICA

AFRICA

GONDWANALAND

300 MILLION YEARS AGO

SIBERIA

LAURENTIA   BALTICA

NORTH AMERICA

CENTRAL EUROPE

BOHEMIA

PANNONIA

GONDWANALAND

AFRICA

SOUTH AMERICA

ARABIA

The Hercynian Orogeny

The formation of the Appalachians was a continuation of the earlier Caledonian Orogeny, and part of a great event in the history of the Earth. At the end of the Paleozoic era, 250 million years ago, almost all of the continents came together to form the super-continent of Pangea. The collisions of the continents caused a series of huge disturbances known as the Hercynian Orogeny (also known as the Alleghenian or Appalachian Orogeny in North America). Though many of the mountains formed during the Hercynian have been eroded or overlain by later events, the Applachians remain as evidence of massive continental collisions.

Smoky Mountains (right)
In North Carolina, the Great Smoky Mountains, part of the Appalachian chain, rise to a height of 2,000 m (6,600 ft) at Clingman's Dome.

THE PRESENT-DAY APPALACHIANS

- first phase of Appalachian formation, including parts of Avalonia
- Piedmont, Ridge and Valley and Blue Ridge Ranges, created by overthrusting after first Appalachian phase
- later ranges in Texas and southern states, caused by collision of South America and North America
- African terrane

— pre–Appalachian coastline

— modern coastlines

Mountains of eastern North America

The mountains of eastern Canada were formed in the Caledonian Orogeny, while further south the formations get progressively younger. This reflects the rotational movement of Gondwanaland in its collision with Laurentia 300 million years ago.

Cross section through Appalachian zone

overthrusting pushes sediment westwards

portion of island arc or micro-continent

sliver of African terrane

Atlantic Ocean

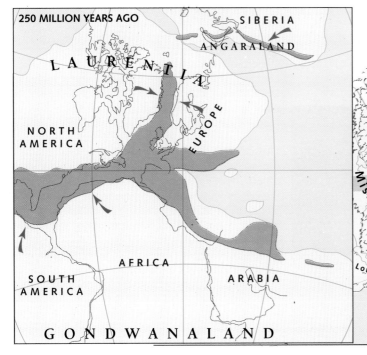

250 MILLION YEARS AGO

SIBERIA

ANGARALAND

LAURENTIA

NORTH AMERICA

EUROPE

AFRICA

ARABIA

SOUTH AMERICA

GONDWANALAND

Metals in the Earth
Sedimentary ores in the Mississippi valley

Our interest in the rocks of the Earth is more than scientific curiosity. The Earth is the source of the minerals that sustain our industrial economies. Humans have been using minerals for thousands of years, but it is only since the Industrial Revolution of the 18th and 19th centuries that we have mined the Earth on a large scale.

Useful metals are present in the Earth in minute proportions and would be impossible to mine economically if they were evenly distributed. Fortunately, geological processes tend to accumulate concentrations of minerals in particular places. Hot mineral-rich liquids, or hydrothermals, can push out from magma chambers into fissures and cracks in surrounding rocks. They may dissolve minerals from their host rock and then cool down and solidify into mineral veins. Alternatively, groundwater can percolate through sedimentary rocks and, over millions of years, leach out certain minerals so that they become concentrated in a layer at the bottom. Knowledge of how these processes work helps economic geologists in the constant search for new mineral sources.

Better technologies of extraction allow lower grades of ore to be used, and supply has kept pace with increasing demand. Recently however, the environmental cost of large-scale mining has become an important issue, particularly in developing countries.

limestone deposits

coastal swamps

UNITED STATES

Appalachians

coastal sediments

ATLANTIC OCEAN

GULF OF MEXICO

THE CREATION OF THE MISSISSIPPI METALS

- lead zinc ore deposits
- limestone
- costal swamps
- coastal sediments
- Appalachian mountains

0   300 km
0   300 miles

N

Galena
The most common ore of lead, (*left*) its chemical composition is lead sulphide, PbS. Galena forms in granular masses, often in medium temperature hydrothermal deposits.

Purple Fluorite (*below*)
Often formed by the action of sulphide rich liquids on

Chalcopyrite (*above*)
Chemically a double suphide of iron and copper (CuFeS2), Chalcopyrite is the principal copper ore in the Mississippi valley.

calcium carbonate rocks, the composition of fluorite is calcium sulphide, CaS.

Ore deposits and the paleoenvironment

Lead, zinc and copper ores occur in significant quantities in sedimentary rocks in the Mississippi valley. A particular set of circumstances arose in the Late Paleozoic which concentrated these metal ores in the region. Ores formed in this way are known as Mississippi valley ores.

About 300 million years ago the Appalachian disturbance was reaching its climax on the eastern margin of North Americ. Hot volcanic magma was being pushed up through the crust, and with it came liquids rich in mineral salts. Meanwhile, in the preceding 50 million years, large amounts of limestone had been deposited in what is now the Missssppi valley.

Some of the hydrothermal ('hot water') liquids were forced out to the west of the Apppalachian zone. They carried copper, zinc and lead, usually as chloride salts. When these liquids met the Mississippi limestones, they reacted with the sulphur in the organic matter of the limestone, and produced insoluble sulphides. It is the sulphides that are the metal ores that are mined today.

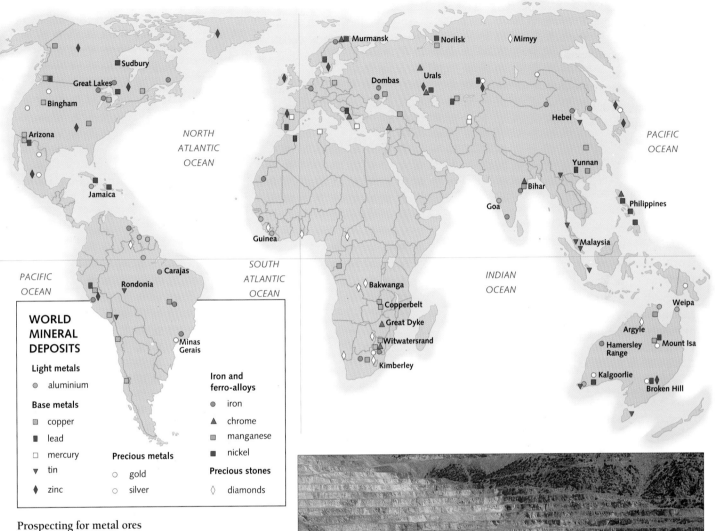

WORLD MINERAL DEPOSITS

Light metals
- aluminium

Base metals
- copper
- lead
- mercury
- tin
- zinc

Precious metals
- gold
- silver

Iron and ferro-alloys
- iron
- chrome
- manganese
- nickel

Precious stones
- diamonds

Prospecting for metal ores

All prospectors use knowledge of the properties of a metal and its ores to locate deposits. At its simplest, panning for gold and silver works because of the density of the native metals. This allows the particles of metal to be carried only so far out of the source vein by running water, before they are deposited, usually in sand and gravel banks, on the inside of bends in streams. Density also lets the gold settle into the bottom of the pan when the silt is washed out. When panners find gold they are not very far away from the mother lode. Minerals found in streams have been almost entirely worked out. More sophisticated technology is now employed to find metals and minerals.

Rock formations containing metals are generally denser than average. Measurements of gravity taken at the surface show up minute variations in the density of rocks buried a considerable distance beneath ground. High gravity readings are taken as an indication of metal-rich rocks, while lower than average can indicate salt deposits, which in turn indicate the possible presence of oil. Magnetic variations are caused by iron ores, and these can also be measured at the surface.

The technique of setting off explosions and then measuring the resulting shock waves in different locations, has been used to analyze the deep interior of the Earth – our knowledge of the boundaries between the crust, the layers of the mantle and the inner and outer core come largely from these seismic studies. Rock formations also show differences in the transmission of seismic waves, and these differences are used to great effect in mineral exploration.

Underlying rock has an effect on surface soils and therefore plant types and varieties. Chemical surveys of soil are a way of finding metal-rich deposits. Surface effects of underlying geology are most obviously shown up on satellite images, and these too are now part of the tools of the exploration geologist.

Finally, the most direct method of testing rocks beneath the surface is to drill boreholes. Analysis of cores from boreholes is used, particularly in the oil industry, to build up a picture of the geology beneath the surface, and to judge the economic value of any deposits.

An open-cast mine
Minerals that are near the surface and occur in low concentrations in their host rock are often mined by open-cast mining. The top layer of soil, known as the overburden, is removed and huge draglines are used to dig out the various ores. Processing of the ores is done on site to save transportation costs. Coal, iron and copper ores are often mined in this way.

The Permian World

The Earth 290 to 250 million years ago

The Permian period is named after the region around the town of Perm in Russia, where the sequence was first recognized. The coming together of the continents into the great landmass, or supercontinent, of Pangea culminated during the Permian. The lowest known sea levels occurred at the end of the Permian – perhaps as much as 100 metres (330 ft) below present levels. At the end of the period there was a series of mass extinctions of marine life forms. This marks the end of the Paleozoic era, and the beginning of a new phase in the history of life on Earth – the Mesozoic or 'middle life' era.

Climate was more strongly zoned than during the Carboniferous. As new mountain chains were formed, differing climatic regions were created within continents. Arid regions bordered those lying under the rain-shadows of upland areas. In equatorial regions the Variscan, Appalachian and Mauranitide mountains blocked the easterly wet equatorial winds, casting a rain shadow over much of central Pangea. Some tropical regions are known to have been subject to monsoon conditions as continental areas emerged above sea level. Desert conditions were widespread in the tropical regions. As northern Europe and mid-North America moved northwards, the equatorial coal swamps of the Carboniferous were replaced by deserts and salt inland seas.

As the new supercontinent of Pangea moved steadily northwards, parts of it remained in south polar latitudes, so that glaciation in the southern continents continued into the Permian. The Gondwana ice cap reached its peak in the early Permian, but then gradually retreated to be replaced by bogs and peat swamps.

Permian rock types

The Permian period is characterized by the continued enlargement of the land area of the Earth. This led to a diminishing of marine sediments and an increase in other types of sediment. The great rock formation of the Permian in the northern hemisphere is known as New Red Sandstone. This was originally thought to be a strong indicator of desert conditions, but it is now known that red sandstone can be formed in a variety of environments. Nevertheless, hot, arid climatic conditions certainly existed over large areas of the continent of Pangea.

The Upper Permian contains more marine formations. Dolomitic limestones – containing magnesium as well as calcium carbonate – and evaporites were formed during repeated incursions of a shallow sea known as the Zechstein Sea across most of Europe. The evaporites are important economic deposits, and were prevalent in the Triassic.

The Permian World
At the start of the Permian period the major continental blocks were edging closer, and during the Permian the final piece was put in place in the make-up of the supercontinent of Pangea. The continent of Angaraland, having rotated through 180°, collided with the eastern margin of Laurentia and Baltica, creating the continent of Laurasia. The Ural Mountains are the result of this collision

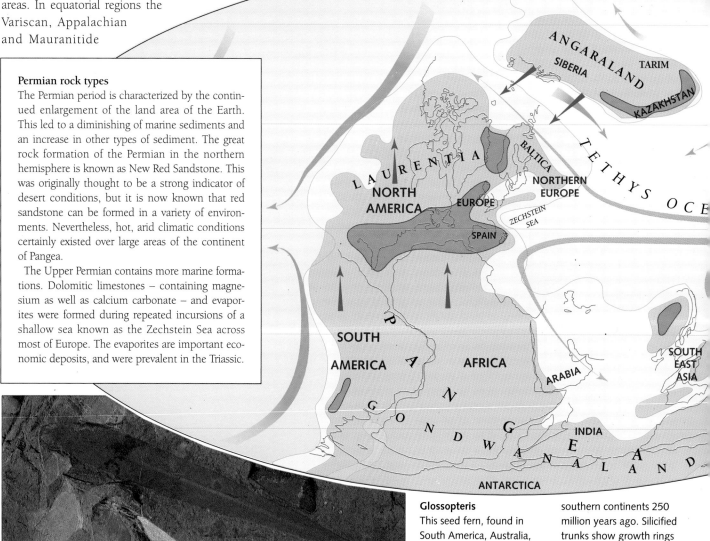

Glossopteris
This seed fern, found in South America, Australia, Africa and India, was the dominant flora of the southern continents 250 million years ago. Silicified trunks show growth rings which indicate a seasonal climate.

The two halves of Pangea, Gondwanaland and Laurasia, were joined in mid-Europe, with the Tethys Ocean stretching away towards the east. The eastern part of Pangea, which eventually went to make up the continent of Asia, may have comprised as many as 11 separate continental plates.

The small continent of Tarim collided with Angaraland during the Late Carboniferous or Early Permian.

Permian Reptiles

The end of the Permian period saw a mass extinction of many forms of marine life, which also affected land plants and animals. Reptiles were able to survive changes in the marine environment; they no longer needed to lay eggs in water, and became much more diverse and widespread on land. During the Permian a group of reptiles – the cynodont therapsids – began to develop warm-blooded characteristics. These were the forerunners of mammals. Because most of the land area of the Earth was linked together, animals tended to be more cosmopolitan. Reptiles spread across Gondwanaland and into China.

Sphenacodon
Reptile from the pelycosaur group. These show some mammal-like characteristics – such as variations in teeth size for different functions.

Pareiasaurus
A 2.5m (8 ft) long reptile. Remains have been found in Permian rocks in Europe.

Scutosaurus
A plant eater similar in size and shape to Pareisaurus.

THE PERMIAN WORLD

	ancient continents
	ancient continental shelf
	ancient mountain chains
	warm ocean currents
	cold ocean currents
BALTICA	ancient place names
	modern coastlines
EUROPE	modern place names
	continental movements

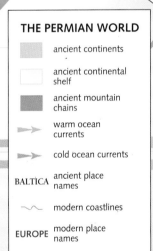

N

AUSTRALIA

Elginia
This spike-headed reptile was about 60 cm (2 ft) long. It was a plant-eater, so presumably the spikes were for defence, or for burrowing.

Titanosuchus
A late Permian reptile with mammal-like characteristics. Probably warm-blooded and with hair replacing the reptilian scales. Found in southern Africa.

Gondwanaland Glaciations in the Permian

The ice ages of the Late Paleozoic

Towards the end of the Carboniferous and into the Permian period, the world's southern continents were in the grip of an intense ice age. The thick ice sheet transported huge quantities of material and deposited it over a vast area. These glacial deposits, known as tills or tillites, are found in present-day South America, Africa, India, Australia and Antarctica. The accumulation of evidence of such widespread glaciation (*see below*) eventually led geologists to conclude that all the southern continents must at one time have been joined in a massive super-continent – Gondwanaland.

The late Paleozoic ice-sheet measured, at its greatest extent, roughly 10,000 by 7,000 kilometres (6,000 by 4,000 miles) (the present Antarctic ice-sheet is about 4,000 kilometres (2,500 miles) in diameter). Continental movement brought Gondwanaland across the South Pole. At this time there was also a notable drop in sea level. By exposing continental shelf areas, this probably led to increased reflection of the Sun's heat (known as albedo) which increased the cooling of the region. Once an ice cap started to form, in the Upper Carboniferous period, this accelerated the albedo process. The supercontinental ice-sheet remained stable for about 30 million years, before rising sea levels and continental movements brought warmer temperatures. By the end of the Permian period the ice-sheet had almost disappeared.

As well as providing evidence of continental drift, these deposits give us indications of the direction of flow of the ice sheets, (through striations, or scratches, which appear on the surface of rocks underlying glacial deposits), the length of the glacial episodes, and the climatic variations in the Upper Carboniferous.

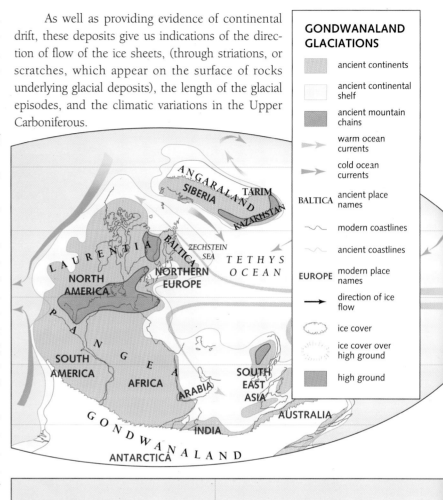

Evidence for Gondwanaland

The evidence accumulated by geologists to support the existence of the super-continent of Gondwanaland comes from a number of sources.

Firstly, the rock strata of all the southern continents are remarkably similar in the late Paleozoic and early Mesozoic periods. The sequence (*far right*) is known as the Gondwana rock succession.

Secondly, rocks of the time contain fossils of plants which show a remarkable similarity. These are plants of the *Glossopteris* fauna which were present on all the southern continents. There is no likely way that the plants could have spread across large bodies of water, and this gives further support to the idea that the continents were joined.

Thirdly, glacial deposits of the same type and age have been found across the southern continents (*map right*). It is possible to use the glacial deposits to map the direction of ancient ice-flows. When these are projected onto a map of reconstructed Gondwana (*top right*) they give a remarkable picture of ice flows radiating out from an area in present-day Antarctica.

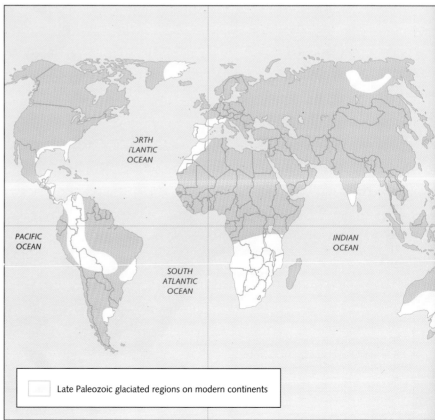

Late Paleozoic glaciated regions on modern continents

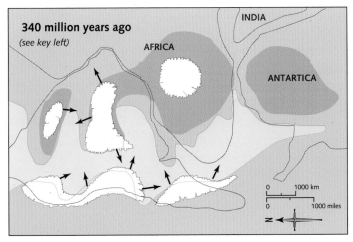

340 million years ago
(see key left)

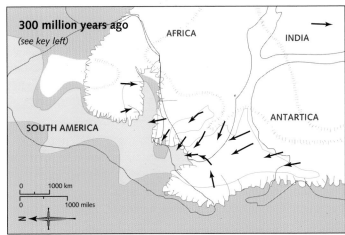

300 million years ago
(see key left)

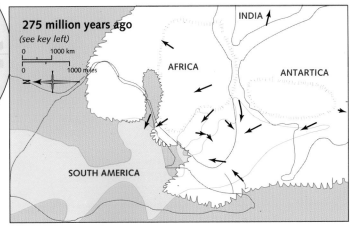

275 million years ago
(see key left)

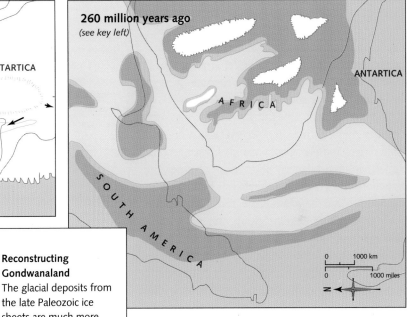

260 million years ago
(see key left)

Cretaceous	Flood basalts Cretaceous in India and South Amercia. Jurassic in South Africa, Antarctica and Australia
Jurassic	Volcanic / basalts Sandstone Volcanics / basalts
Triassic	Sandstone including red-beds Comglomerates Some continental sandstone
Permian	Glacial tillites interbedded with coal seems and sandy shales
Upper Carboniferous	Coal and tillites resting on basement volcanics

Reconstructing Gondwanaland

The glacial deposits from the late Paleozoic ice sheets are much more widespread than those of earlier glaciations. Tillites are found on every continent in the southern hemisphere (*left*). Striations on rock surfaces can be used to give clues as to the direction of ice flows, and this information is used to reconstruct Gondwanaland (*top left*).

At its greatest extent the Gondwanaland ice cap reached present-day Brazil, central Africa, southern India and the southern half of Australia.

Gondwanaland glaciations

340 million years ago
In the Lower Carboniferous local ice caps had formed on areas of high ground across South America and Africa, and on an island arc to the west. An inlet of continental shelf sea covered much of South America.

275 million years ago
In the Early Permian the ice sheets were beginning to retreat back to high ground. This caused a global rise in sea levels which in turn brought more warm water to the region, flooding low-lying areas.

300 million years ago
The end of the Carboniferous saw a massive ice sheet was covering most of the continents in the region, and a substantial portion of the continental shelf sea. The influx of warmer water via ocean currents limited the extent of the ice sheet.

260 million years ago
Local ice caps, restricted to a few areas of high ground, were all that remained by the end of the Lower Permian. The ice sheet lasted longer elsewhere on Gondwanaland.

Mass extinctions of Marine Life
The end of the Paleozoic era

The evolution of life involves not only the emergence and spread of species but also extinction. A species of animal or plants tends to increase in number as long as there are no external controls on the growth of population. Eventually the increasing numbers will put a strain on the food supply or other aspects of the environment in which the organisms live. This can be relieved by the organisms spreading to new places or adapting to new environments. But within the original environment some individuals may, through the process of accidental mutation, evolve more effective ways of, for example, feeding or reproducing. The numbers of these individuals will then increase, and their mutations will be reinforced by breeding between them. This group will form a new species and will replace their predecessors. This form of extinction is a local process, and is not dependent on any change in environment – it is the relentless pressure of numbers that causes the adaptation and extinction.

But at certain times in the Earth's history a surprisingly large numbers of species, including entire orders and classes of organisms, have become extinct, within a relatively short space of time. The extinction of a group of animals can be caused by competition for the same ecological niche from a newly emerged or migrant group. But such large-scale extinctions are almost certainly the result of widespread environmental changes. Some organisms inevitably adapt better than others, and it is often the dominant animals, who have benefited from a high degree of specialization, which are unable to adapt when conditions are no longer suitable.

There are five significant mass extinctions in the Earth's history: at the end of the Ordovician (440 million years ago), Devonian (350 million years ago), Permian (250 million years ago), Triassic (210 million years ago) and Cretaceous (65 million years ago) periods. Because the geological time-scale was originally divided according to the presence of fossils in the rocks, most of these mass extinctions coincide with the end of major geological periods. All these extinctions involved marine life, and the three most recent affected land animals as well. The most important were at the end of the Permian period, which is also the end of the great Paleozoic era, and at the end of the Cretaceous, which was the close of the Mesozoic era and brought about the demise of the dinosaurs.

The causes of these periodic mass extinctions are uncertain. Understanding what wiped out the dinosaurs or decimated Paleozoic life forms remains one of the great puzzles. Extraterrestrial causes like comets are dramatic explanations. Comets and asteroids do occasionally collide with the Earth, and the impact of a comet 10 kilometres (6 miles) in diameter would produce dust clouds and extensive forest fires, which would slow photosynthesis and induce extinction down the animal food chain. More humdrum explanations like climatic change have taken on a new significance in our own time. The uniting factor is environmental change. The more specialized an organism is, the less likely it is to adapt and survive.

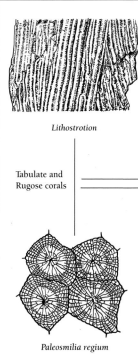

Lithostrotion

Tabulate and
Rugose corals

Paleosmilia regium

Mass extinctions

Many groups of marine invertebrates that are important fossils in Paleozoic rocks became extinct at the end of the era. Half of all marine invertebrate families disappeared. Those that survived were also severely affected – 90 per cent of all species are thought to have become extinct.

Some groups, like the trilobites, had been in decline for some time but other disappeared fairly suddenly. Groups that became extinct include rugose and tabulate corals, productid brachiopods, goniatites and trilobites, *(illustrations, right)*.

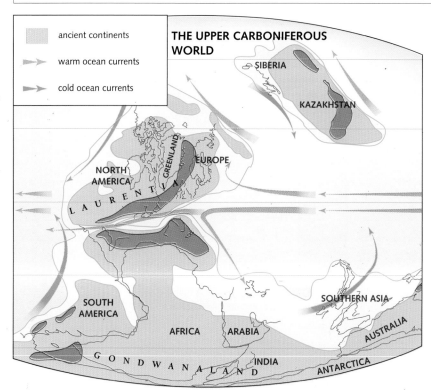

Changes in ocean currents at the end of the Paleozoic era
The configuration of the continents has a profound effect on the world climate. Ocean currents are the major determinants of climate and these can obviously only flow around the landmasses. The major continental change at the end of the Paleozoic era was the collision of Africa and South America with North America and Europe. This major cataclysmic event lasted for several million years, creating mountain chains, earthquakes and volcanic activity. At some time during this collision of landmasses the channel through which the major warm ocean currents flowed became closed. The currents then began to

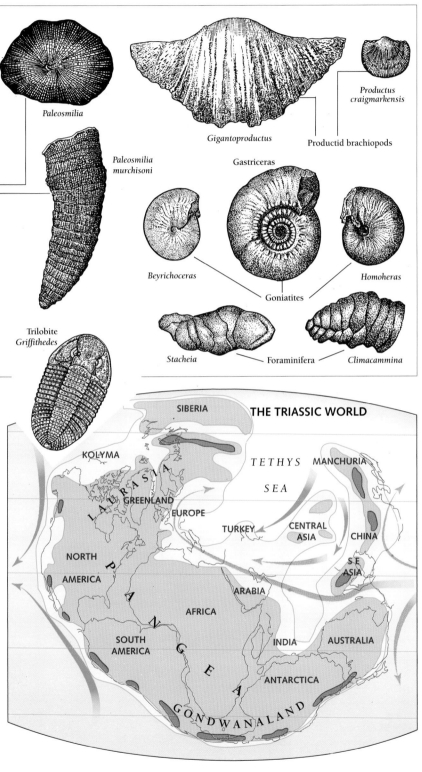

Paleosmilia

Paleosmilia
murchisoni

Gigantoproductus

Productus
craigmarkensis

Productid brachiopods

Gastriceras

Beyrichoceras

Goniatites

Homoheras

Trilobite
Griffithedes

Stacheia

Foraminifera

Climacammina

THE TRIASSIC WORLD

SIBERIA

KOLYMA

TETHYS

MANCHURIA

SEA

LAURASIA

GREENLAND

EUROPE

TURKEY

CENTRAL
ASIA

CHINA

NORTH
AMERICA

S E
ASIA

PANGEA

ARABIA

AFRICA

SOUTH
AMERICA

INDIA

AUSTRALIA

ANTARCTICA

GONDWANALAND

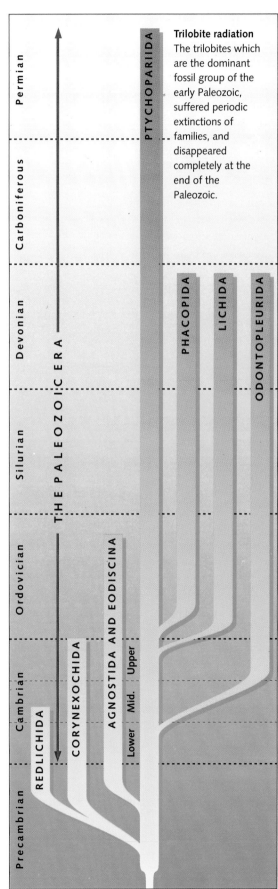

Trilobite radiation
The trilobites which are the dominant fossil group of the early Paleozoic, suffered periodic extinctions of families, and disappeared completely at the end of the Paleozoic.

flow in entirely different directions. Although the continents moved together very slowly the change in ocean currents would have been very sudden and would have affected all the other ocean currents. This in turn would have changed the Earth's climate and weather patterns completely, possibly within a single year. The effect on life forms would obviously have been immense. Though it is tempting to believe that dramatic extinction events must have extraterrestial causes such as comets, it is entirely possible that major changes in the world's ocean currents hold the answer.

Europe and Asia Joined
The Ural Mountains and the Siberian Traps

The Urals are generally seen as the geographical land barrier between the continents of Asia and Europe. This view reflects exactly the geological history of the two continents. For most of the Earth's history Siberia has been a separate continent. Siberia completed its long journey away from Gondwanaland about 250 million years ago, and collided with the eastern side of the continent of Europe – then part of Laurentia. As well as causing the massive upheaval that is evident in the Ural mountain belt, Siberia was the last major piece of continental crust to join the great super-continent of Pangéa, which comprised all the continents of the Earth.

Siberia is an ancient craton surrounded by younger zones of rock, formed during different orogenic episodes over the past 600 million years. There are fossils of marine animals from the Proterozoic period on the edges of the craton – at times during the Cambrian period 550 million years ago the continent was flooded by shallow seas. Later in the early part of the Paleozoic, around 450 million years ago, its southern margin underwent intense volcanic activity. During the Carboniferous period coal-forming swamps covered large parts of the margins and central craton of Siberia. These, together with the swamps covering parts of northern Europe and North America may have resulted from the huge amounts of sediments being washed down off the newly formed mountains in the Hercynian and Ural regions, creating flat areas with slow-flowing rivers.

Siberia was in collision with the smaller Kazakh plate before the collision of Siberia with Laurentia which coincided with the formation of the Siberian Traps. This vast area of sulphur-rich lava was poured onto the surface of the continent 250 million years ago.

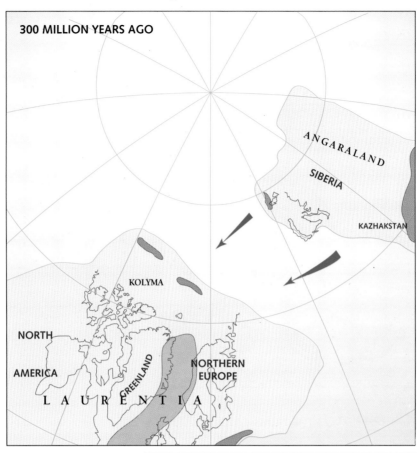

300 MILLION YEARS AGO

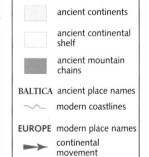

300 million years ago
In the Upper Carboniferous the small plate carrying present-day Kazakhstan collided with the more massive Siberian plate, causing mountain-building. The new combined continent is known as Angaraland. This was located to the north and east of the continent of Laurentia, which comprised northern Europe, Greenland and North America, with the small continent of Kolyma attached to the northern margin.

	ancient continents
	ancient continental shelf
	ancient mountain chains
BALTICA	ancient place names
~	modern coastlines
EUROPE	modern place names
→	continental movement

250 MILLION YEARS AGO

250 million years ago
At the end of the Paleozoic era the Siberia/Kazakh plate (also known as Angaraland) continued its drift south-west, and collided with the north-east margin of Laurentia. To the south the supercontinent of Gondwanaland was in collision with Laurentia.

North Urals

The Ural Mountains stretch for over 2,000 kilometres (1,250 miles), almost directly south from the northern coast of Russia to Kazakhstan. They vary from the sharp peaks seen here, to low rounded hills, though they form a definite boundary between the west Siberian plain and the European plain. The highest parts of the Urals are around 1.600 metres (5,250 ft) compared with the 5,500 metres (18,000 ft) of the much younger Caucasus Mountains.

The Urals have been eroded over time, but also the lack of plate movements in the region has caused the mountains to gradually subside.

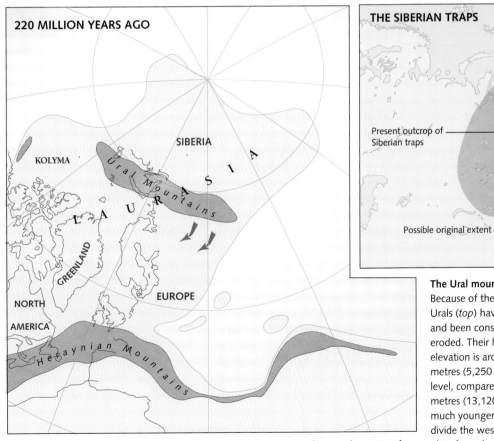

220 MILLION YEARS AGO

KOLYMA

SIBERIA

Ural Mountains

LAURASIA

GREENLAND

NORTH AMERICA

EUROPE

Hercynian Mountains

THE SIBERIAN TRAPS

Present outcrop of Siberian traps

Possible original extent

The next 220 million years
The Angara plate rotated clockwise bringing it into collision with the entire eastern edge of Laurentia. This collision threw up the Ural Mountains. The rest of the Asian continent is made up of a complex series of small plates which gradually became joined on to the Angara plate.

The Ural mountains
Because of their age, the Urals (*top*) have subsided and been considerably eroded. Their highest elevation is around 1,600 metres (5,250 ft) above sea level, compared to 4,000 metres (13,120 ft) of the much younger Alps. They divide the west Siberian plain from the European plain – an orogenic zone lying between two old continental cratons.

The Siberian traps (*above*)
This huge area of lava flows was formed at about the same time as the Urals. Over about half a million years thousands of cubic kilometres of lava were erupted onto the surface. There is speculation that the effects of the Siberian traps on the global climate might have been enough to cause the mass extinctions at the end of the Paleozoic.

Part Five
The Mesozoic Era

The Triassic World
The Earth 250 to 210 million years ago

The Triassic (named after a three-fold rock formation in Germany) is the first period of the Mesozoic era. It is a difficult period for geologists to study, because there are relatively few marine sediments, and shell fossils are rare. This is probably because Pangea, the landmass into which all the major continents were locked, remained above sea level. The area of continental shelf seas, the most productive sites for marine fossil preservation, was severely reduced. Instead this was a time of continental deposits – red sandstones were deposited by rivers and in deserts. Evaporites were formed from dried out salt flats and lakes.

After the great extinctions at the end of the Paleozoic era the remaining life-forms had the opportunity to diversify and increase in abundance. Although there is a relative scarcity of marine sediments from the Triassic period, significant events in the development of land life occurred.

The first dinosaurs, the reptiles that were to dominate the later Jurassic and Cretaceous periods, arose during the Triassic, as did the first crocodiles, ichthyosaurs and turtles. The Triassic also saw the emergence of the first mammals – though for the first 150 million years of their existence they were of minor importance compared to the great reptiles. Triassic forests were dominated by conifers, cycads (palm-like trees) and gingkos ('maidenhair trees'). Early conifers from the Triassic period have been preserved in the Petrified Forest of Arizona.

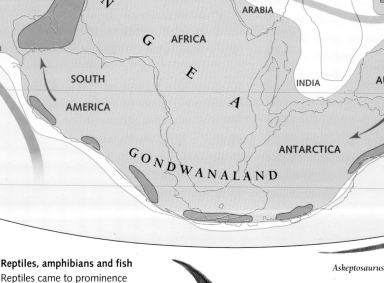

Kuehneosaurus
Strongly resembling the present-day gliding lizards of the tropics, fossilized remains of *Kuehneosaurus* have been found in caves in Wales and southern England. Long ribs protruded from either side of the body and were covered by skin, enabling the lizard to glide from tree to tree.

THE TRIASSIC WORLD

- ancient continents
- ancient continental shelf
- ancient mountain chains
- warm ocean currents
- cold ocean currents
- continental movements
- BALTICA ancient place names
- modern coastlines
- EUROPE modern place names

SIBERIA

KOLYMA

GREENLAND

EUROPE

TETHYS OCEAN

MANCHURIA

TURKEY

CENTRAL ASIA

CHI

SPAIN ITALY

NORTH AMERICA

S E ASIA

N

ARABIA

G

AFRICA

E

INDIA

AUSTRALIA

SOUTH AMERICA

A

ANTARCTICA

GONDWANALAND

Reptiles, amphibians and fish
Reptiles came to prominence in the hot dry conditions of the Permian and Triassic periods. Though freed of dependence on water by their ability to lay amniotic eggs, some reptiles returned to marine environments, as ecological opportunites arose.
Fish were little affected by the mass extinction at the end of the Permian period, and thrived in the Triassic.

Askeptosaurus

Mixosaurus

The position and movement of the continents

For most of the Triassic period the continental cratons of the Earth were locked into the supercontinent of Pangea (which means 'all of the Earth'). The break up of Pangea began at the end of the Triassic period 210 million years ago, with the eruption of basalt along the margins between present-day North America and north-west Africa. The initial trigger for this break-up was the formation of a series of domes, presumably caused by upswelling from the mantle. Where these coincided with lines of weakness (old continental margins) there was splitting followed by the out-flowing of massive quantites of basalt. These flows of basalt spread out over the continents at the end of the Triassic and beginning of the Jurassic period 200 million years ago.

The clockwise rotation of Pangea continued, taking most of eastern Gondwana (Antarctica, Australia and India) into southerly, polar latitudes, while the western part, including South America, Africa and Arabia moved gradually northwards towards the equator. This movement created disturbances and subduction zones along the whole southwestern margin of Pangea – causing the mountain-building seen on the map.

Rhamphorhynchus
An early pterosaur, with wings which were stretched over a long fourth 'finger'.

The long tail was used for steering. The *Rhamphorhynchus* was the size of a seagull.

Megazostrodon
The best known of the first true mammals, *Megazostrodon*, was roughly the shape and size of a shrew. An insect-eater, fossils have been found in the Late Triassic rocks of South Africa.

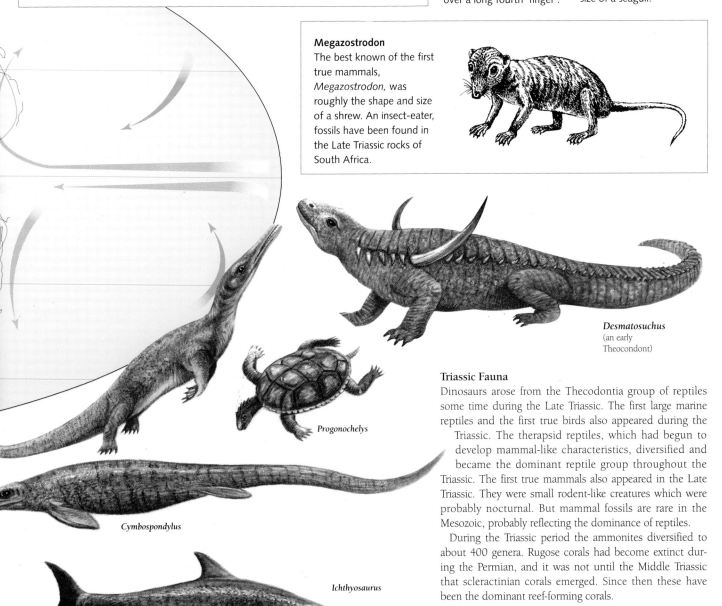

Desmatosuchus
(an early
Theocondont)

Progonochelys

Cymbospondylus

Ichthyosaurus

Triassic Fauna

Dinosaurs arose from the Thecodontia group of reptiles some time during the Late Triassic. The first large marine reptiles and the first true birds also appeared during the Triassic. The therapsid reptiles, which had begun to develop mammal-like characteristics, diversified and became the dominant reptile group throughout the Triassic. The first true mammals also appeared in the Late Triassic. They were small rodent-like creatures which were probably nocturnal. But mammal fossils are rare in the Mesozoic, probably reflecting the dominance of reptiles.

During the Triassic period the ammonites diversified to about 400 genera. Rugose corals had become extinct during the Permian, and it was not until the Middle Triassic that scleractinian corals emerged. Since then these have been the dominant reef-forming corals.

Desert Sandstone

The northern continents turn hot and dry

In the Late Permian, period 250 million years ago, the southern continents were located around the South Pole, while the northern parts of Pangea (Laurentia, Baltica and parts of Asia) were located in the tropics. The northern continents became warm and dry, deserts formed across large parts of what is now North America, Europe and northern Asia.

We know from our present world that special circumstances are needed to produce the extremely arid conditions in which deserts form. In particular, prevailing winds must be prevented from carrying rainwater from the oceans, either by atmospheric conditions or by mountain ranges creating rain shadows. The necessary high stationary atmospheric pressure is found at about 30° either side of the equator. In the Permian and Triassic periods the lack of rain particularly affected areas in the rain shadow of new mountain chains like the Appalachians. The geography of the continents once again profoundly affected conditions on Earth.

The deserts of the Permian and Triassic periods lasted for varying amounts of time as local conditions fluctuated. But they persisted for long enough to form huge areas of sandstone. Whereas most sandstones, like almost all other sedimentary rocks, are deposited under water, desert sandstones are formed on land. The lack of water in deserts means that rocks are eroded by wind action. The resulting material is blown around on prevailing winds. It gets buried beneath more and more sand, and eventually lithifies into rock under pressure from above. Desert sandstones are characterized by smooth spherical grains, caused by reworking by the wind, and loose structure with large spaces between the grains. This makes them perfect as a reservoir rock for oil (*see page 104*).

Wind-polished millet seed
Erosion in extremely arid conditions is by wind. This causes the rounding of individual sand grains.

'Millet seed' grains are highly polished grains of quartz sand, blown loosely across hardened surfaces.

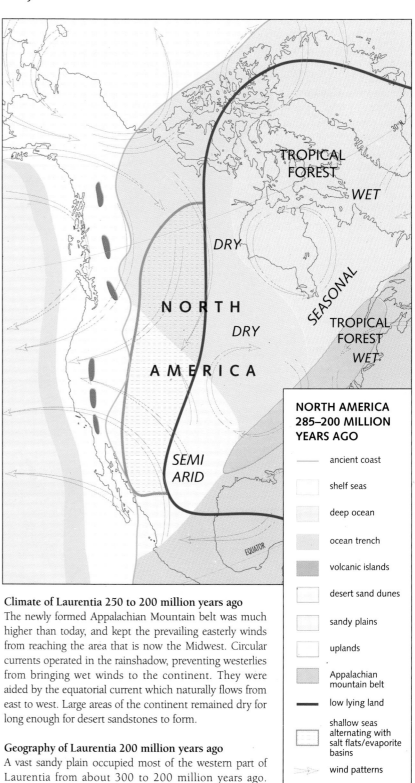

TROPICAL FOREST

WET

DRY

NORTH

DRY

AMERICA

SEASONAL

TROPICAL FOREST WET

SEMI ARID

EQUATOR

NORTH AMERICA 285–200 MILLION YEARS AGO

- ——— ancient coast
- shelf seas
- deep ocean
- ocean trench
- volcanic islands
- desert sand dunes
- sandy plains
- uplands
- Appalachian mountain belt
- —— low lying land
- shallow seas alternating with salt flats/evaporite basins
- ⇢ wind patterns

Climate of Laurentia 250 to 200 million years ago
The newly formed Appalachian Mountain belt was much higher than today, and kept the prevailing easterly winds from reaching the area that is now the Midwest. Circular currents operated in the rainshadow, preventing westerlies from bringing wet winds to the continent. They were aided by the equatorial current which naturally flows from east to west. Large areas of the continent remained dry for long enough for desert sandstones to form.

Geography of Laurentia 200 million years ago
A vast sandy plain occupied most of the western part of Laurentia from about 300 to 200 million years ago. Changing sea levels brought the coastline further east and back west again. This, together with erosion of the uplands to the east and south, recycled this great quantity of sand over and over again.

At the beginning of the Jurassic period 200 million years ago, arid conditions in the south-west, combined with onshore northerly winds led to the deposition of a vast quantity (40,000 cubic kilometres) of well-graded sand along the western margin of the continent, with great dunes forming in the south-west. The remnants of these dunes are seen today in the Navajo Desert of Utah.

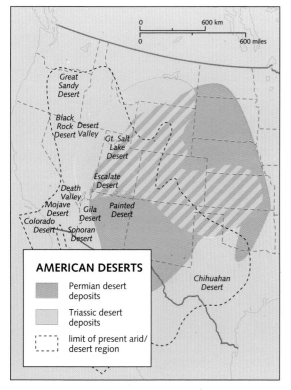

AMERICAN DESERTS

- Permian desert deposits
- Triassic desert deposits
- limit of present arid/ desert region

The climate and deserts of North America

The dune deserts of the Permian, Triassic and Jurassic periods bordered the ancient west coast of the continent. Today the coastline is much further west, due to the elevation of the Rockies and the accretion of new material onto the western edge of the continent. The arid lands now lie on a plateau between the rain shadows of mountain ranges to the west and east. Pacific westerlies drop their rain over the coastal ranges, while warm wet southeasterlies from the Gulf of Mexico bring rain only as far as the Rockies.

The world's deserts

Most deserts lie in the tropical regions, but not at the equator. The Sahara and Arabian deserts lie on the Tropic of Cancer, the Great Australian and Kalahari deserts on the Tropic of Capricorn. These areas have stationary high atmospheric pressure, and negligible rainfall – less than 1 cm (.4 in) per year in some areas. Deserts outside the tropics have low rainfall usually because of their geographical position – either a long way from rain-giving sea, or in the rain shadow of mountain ranges. Antarctica is a desert since it has very low rainfall due to its own high pressure cell.

Northern Sahara (left)

The Akle dunes from the northern Sahara at Erg Bourahet in eastern Algeria, are formed from loose sand, subject to high winds in arid conditions. The dunes are extremely mobile – when winds reach certain speeds sand is blown over the top of the ridge, and then moves down the far side.

Weathering pillars (left)

These landforms in Arches National Park in the United States are the remnants of erosion by rivers earlier in the region's history. Deserts of this type are often subject to flooding or invasion by seas, which wash away the loose wind-blown material and dump it in valleys, lakes and other depressions.

Monument Valley (left)

The spectacular scenery of Monument Valley in Utah, United States, is the result of erosion of rocks of differing hardness. The great buttes are remnants of water, rather than wind, erosion.

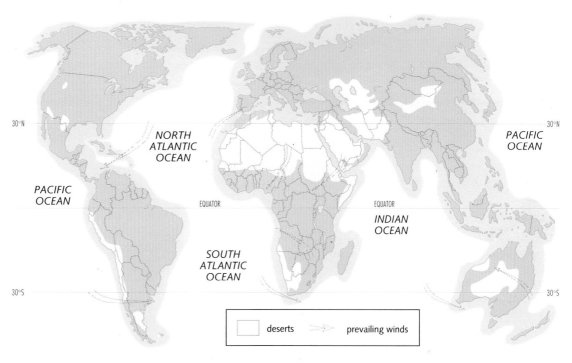

deserts   prevailing winds

Reptiles and Birds
Dominant animals of the Mesozoic

Amphibians, which are the precursors of reptiles, are tied to seas and rivers by their need to lay eggs in water. The crucial stage in the development of reptiles was the ability to lay eggs on land. Once free of dependence on water, reptiles could spread rapidly. At the end of the Carboniferous the climate of much of the world became hotter and drier. This environment was unsuitable for amphibians, and reptiles thrived at their expense. Reptiles became the dominant land animals from the Permian period until the mass extinction at the end of the Cretaceous – a period of over 200 million years.

Having emerged in the late Jurassic, birds had become widespread and highly specialized by the end of the Cretaceous 75 million years later. Birds are direct descendants of dinosaurs and in some ways are the living remnants of those great reptiles. When the dinosaurs died out, and before large mammals had developed, huge numbers of new birds evolved – some of which were extremely exotic. Large meat-eating birds, like *Diatryma* which stood over 2 metres (6 ft) tall, preyed on small mammals. The largest known flying bird was the vulture *Argentavis* – with a wingspan of over 7 metres, while the seabird *Osteodontornis* had a wingspan of over 5 metres (16 ft).

Birds, unlike their reptile ancestors, are warm-blooded. Their evolution in the Tertiary period and up to the present has been as spectacular as that of their fellow warm-blooded animals – the mammals.

Hesperornis
Fossilized remains of this Cretaceous bird have been found in limestone deposits in North America. At 1.7 m (5.5 ft) this flightless diving bird bore a strong resemblance to a modern cormorant.

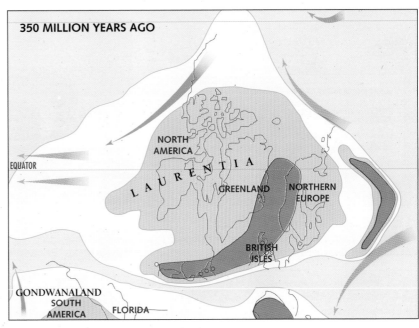

THE FIRST REPTILES

- ancient continents
- ancient oceans
- evidence of tetrapods
- modern coastlines

350 MILLION YEARS AGO

EQUATOR

NORTH AMERICA

LAURENTIA

GREENLAND

NORTHERN EUROPE

BRITISH ISLES

GONDWANALAND
SOUTH AMERICA

FLORIDA

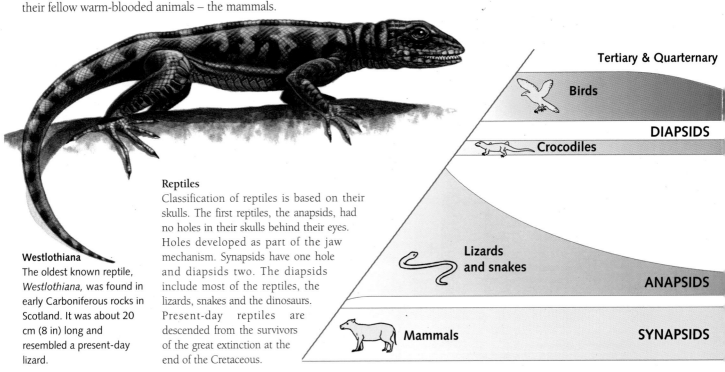

Westlothiana
The oldest known reptile, *Westlothiana*, was found in early Carboniferous rocks in Scotland. It was about 20 cm (8 in) long and resembled a present-day lizard.

Reptiles
Classification of reptiles is based on their skulls. The first reptiles, the anapsids, had no holes in their skulls behind their eyes. Holes developed as part of the jaw mechanism. Synapsids have one hole and diapsids two. The diapsids include most of the reptiles, the lizards, snakes and the dinosaurs. Present-day reptiles are descended from the survivors of the great extinction at the end of the Cretaceous.

Tertiary & Quarternary

Birds

DIAPSIDS

Crocodiles

Lizards and snakes

ANAPSIDS

Mammals

SYNAPSIDS

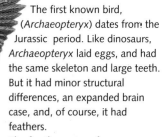

The first known bird, (*Archaeopteryx*) dates from the Jurassic period. Like dinosaurs, *Archaeopteryx* laid eggs, and had the same skeleton and large teeth. But it had minor structural differences, an expanded brain case, and, of course, it had feathers.

The fossil remains of *Archaeopteryx* (*below*) were found in late Jurassic limestone in Solenhofen, Bavaria.

Early tetrapods

The first four-legged creatures (tetrapods) were amphibians living on land, but dependent on water. The maps (*below*) show how some of the early finds are grouped around the margins of the continent of Laurasia. Over the next 50 million years the tetrapods moved inland.

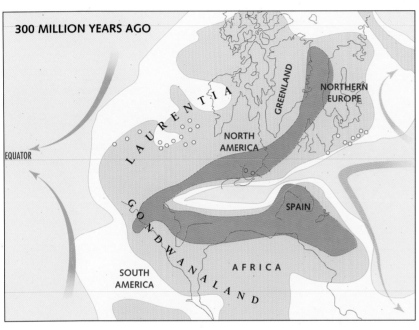

300 MILLION YEARS AGO

EQUATOR

LAURENTIA
GREENLAND
NORTHERN EUROPE
NORTH AMERICA
SPAIN
GONDWANALAND
SOUTH AMERICA
AFRICA

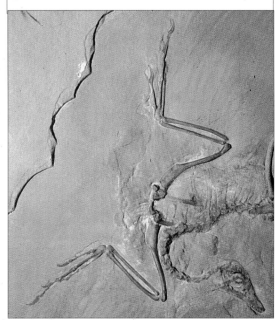

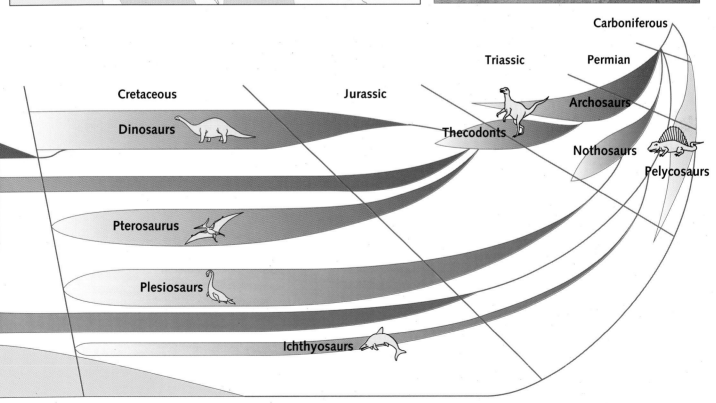

Carboniferous

Triassic   Permian

Cretaceous   Jurassic   Archosaurs

Dinosaurs

Thecodonts

Nothosaurs

Pelycosaurs

Pterosaurus

Plesiosaurs

Ichthyosaurs

The Jurassic World

The Earth 250 to 145 million years ago

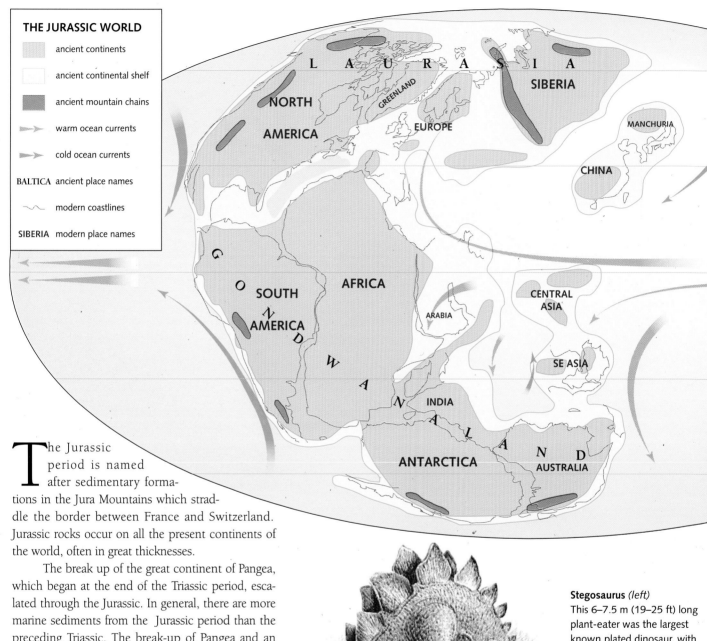

T he Jurassic period is named after sedimentary forma-tions in the Jura Mountains which strad-dle the border between France and Switzerland. Jurassic rocks occur on all the present continents of the world, often in great thicknesses.

The break up of the great continent of Pangea, which began at the end of the Triassic period, esca-lated through the Jurassic. In general, there are more marine sediments from the Jurassic period than the preceding Triassic. The break-up of Pangea and an overall rise in sea levels were the probable cause. Marine sediments around the edges of the North and Central Atlantic show that this ocean was begin-ning to form at the time. Parts of the Indian Ocean also date from this era, as the pieces of continent that were to form southern and Southeast Asia were moving away from Gondwanaland – a continent that had existed since Precambrian times.

The dominant land animals of the Jurassic period are the dinosaurs. The most important fossils used for dating the rocks of the Jurassic are the ammonites. Free-floating ammonites dispersed rapidly over wide areas, and show rapid changes in shell structure.

Stegosaurus *(left)*
This 6–7.5 m (19–25 ft) long plant-eater was the largest known plated dinosaur, with a double row of bony plates along the neck and back, and two pairs of tail spikes.

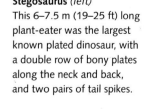

Megalosaurus *(left)*
This 9 m (30 ft) predator, stood on two powerful legs, with saw-edged teeth, muscular arms and toe and finger claws.

The position of the continents

The first real splitting of the great southern continent of Gondwanaland began during the Jurassic period. The parts of Gondwanaland that were to become joined to Asia began to split away from its eastern edge.

As Africa and South America rotated anti-clockwise and split away from North America, the Tethys Ocean began to close.

Rifts between North America and Europe and in the North Sea – the prelude to the opening of the North Atlantic – also appeared at this time.

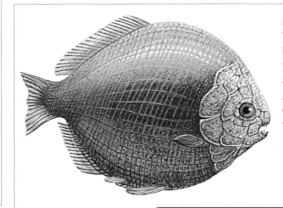

Dapedium (*left*)
This 30 cm (1 ft)-long fish lived near the shores. The thick scales, which provided armour plating, as well as good supplies of sediment ensured that fossils were well preserved.

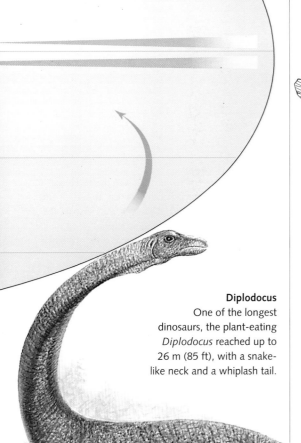

Ginkgo huttoni
The ginko or ginkgo family emerged in the Carboniferous period, and reached their greatest abundance in the Triassic and Jurassic periods.

Pentacrinitus
This crinoid (*above*) is preserved in Jurassic rock. Limestones of this period are often composed entirely of shells of marine animals.

Gymnosperms
The Jurassic period was the last to be dominated by the gymnosperms, or seed-bearing plants such as *Sagenopteris* (*right*) and *Coniopteris* (*left*).

Sagenopteris

Diplodocus
One of the longest dinosaurs, the plant-eating *Diplodocus* reached up to 26 m (85 ft), with a snake-like neck and a whiplash tail.

Coniopteris

Jurassic Flora and Fauna

Reptiles, the dominant group, occurred on sea and land and even in the air, as flying forms (pterosaurs) developed. Dinosaurs reached their maximum size during the Jurassic period. Birds became more widespread.

Ammonites were abundant and are the dominant marine fossils together with belemnites – another family of cephalopods. Brachiopods and echinoids continued to be abundant. Many gastropods appeared at this time.

The Jurassic saw the first appearance of wasps and the first known moths also developed.

The Opening of the Central Atlantic
North America separates from Africa and South America

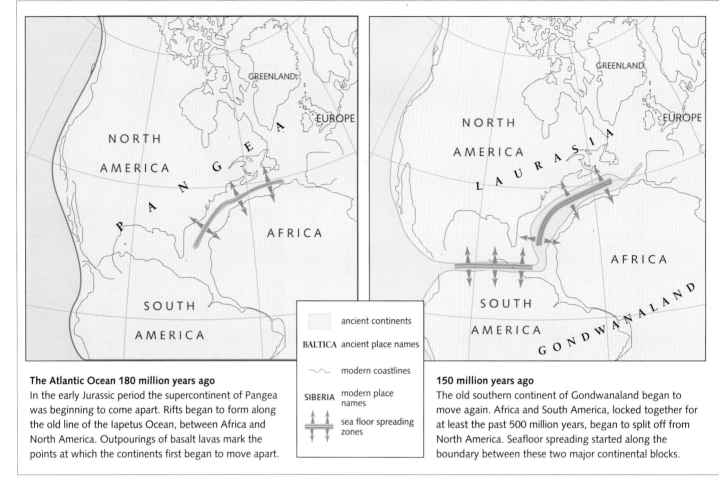

The Atlantic Ocean 180 million years ago
In the early Jurassic period the supercontinent of Pangea was beginning to come apart. Rifts began to form along the old line of the Iapetus Ocean, between Africa and North America. Outpourings of basalt lavas mark the points at which the continents first began to move apart.

150 million years ago
The old southern continent of Gondwanaland began to move again. Africa and South America, locked together for at least the past 500 million years, began to split off from North America. Seafloor spreading started along the boundary between these two major continental blocks.

Legend:
- ancient continents
- **BALTICA** ancient place names
- modern coastlines
- **SIBERIA** modern place names
- sea floor spreading zones

During the earliest part of the Mesozoic era, 250 million years ago, the supercontinent of Pangea reached its greatest extent. From then on, continents began to break away from the great landmasses that had formed over the previous 1,000 million years.

Early in this break-up the ancient southern continent of Gondwanaland split off from the northern landmass of Laurasia (comprising present-day North America, northern and central Europe and part of Asia). Africa and South America remained as one landmass and moved south-east, while North America and Europe moved north-west. In between the central part of the Atlantic Ocean was born.

The opening of this ocean basin was a repetition of much earlier events. The Iapetus Ocean was a previous version of the Atlantic, and its closure had led to the disturbances which built the Appalachian Mountains. The continents split apart along roughly the same line, except that Florida and a sliver of the old African continent were left behind as a strip of coastal plain on the eastern margin of North America.

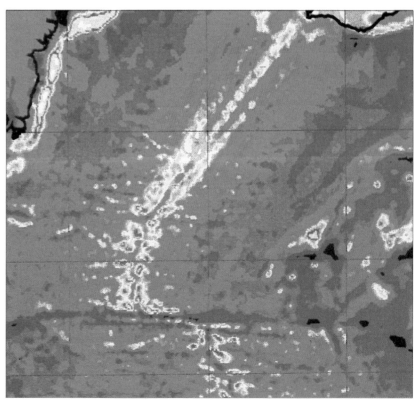

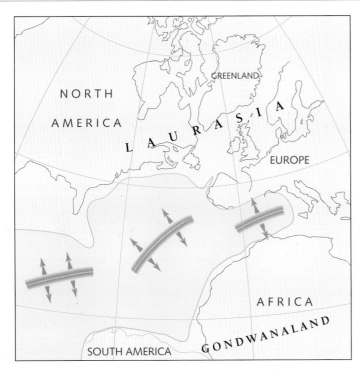

120 million years ago
By the early Cretaceous period a wide ocean separated
North America from Africa and South America. Southern
Europe remained attached to the northern margin of Africa,
as Africa swung anti-clockwise. A new ocean basin was
created in the Gulf of Mexico.

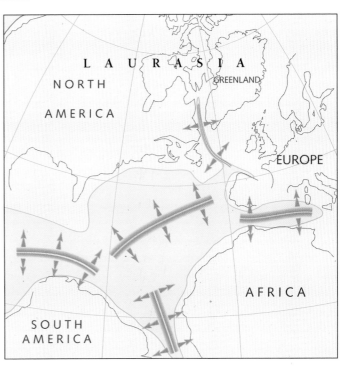

85 million years ago
The width of the central Atlantic stabilized as other
continental movements began to occur. South America
and Africa were about to begin their separation, while an
extension of the Atlantic would temporarily open up
between North America and Europe.

The Central Atlantic
Ocean basins are created by
new igneous material being
poured out onto the sea
floor along a central fissure,
and then pushed sideways
by currents in the underlying
mantle. The central rift
develops into a ridge as
molten rock from the mantle
solidifies to form chains of
volcanic 'mountains'. As
these ridges are successively
pushed aside a characteristic
topography develops as
shown on this physiographic
'reconstruction' (right).

Mid-Atlantic Ridge
The central ridge (left) now
stretches from the Arctic
basin to the Antarctic
Ocean. Flat abyssal plains
stretch away on either side,
getting progressively older
until they meet the
continental shelf edges.

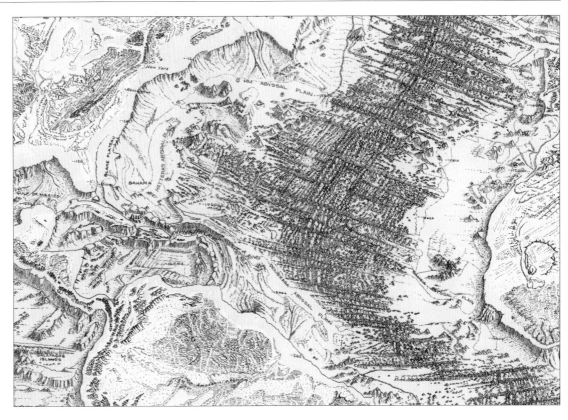

Ammonites

Variety and abundance

The ammonites of the Mesozoic era are instantly recognizable, and for many people they are the most beautiful creatures preserved in the fossil record. They are also crucially important in dating rocks of the Mesozoic, and are indicators of ancient maritime conditions.

Ammonites were free-floating, using their many-chambered shells to adjust their buoyancy, living mainly in shallow marine and coastal waters. They ranged from 1 cm to 2 metres (.4in to 6.5 ft) in diameter. They became extinct 65 million years ago, at the same time as the dinosaurs - the only surviving close relative is the Nautilus.

Individual groups and species of ammonite were acutely sensitive to water temperature, depth and salinity. Different groups would live at varying depths, at different distances from the shore, nearer or closer to estuaries and so on. Knowledge of the environmental limits of the different groups is used to reconstruct the marine environments of the time.

Ammonite fossil finds from part of the lower Jurassic period, are shown (*below*) plotted onto a map of the western Tethys Ocean 200 million years ago. As ammonites adapted and changed quickly, it is possible to be very precise about the time when certain groups of species occurred. This makes them the perfect zone fossil. The ammonite sites are grouped along the northern shore of the ancient Tethys Ocean. Some are found at the margin of the continental shelf, while other groups are found along the coastlines. All groups were restricted to the shallow waters of the continental shelf.

Plotting sites where fossils of different ammonites groups have been found onto the geography of the time when they lived, shows that there was a strictly defined southern limit to the movement of one group of ammonites (Euro-Boreal), while other groups had a northern limit (Mediterranean). Modern coastlines are shown in their ancient positions.

Ammonites in limestone
This rich cluster of fossilized ammonites was found in Jurassic limestone at Robin Hood's Bay on the eastern coast of England. At the time when the rock was formed, two hundred million years ago, this region was covered in warm shallow tropical seas.

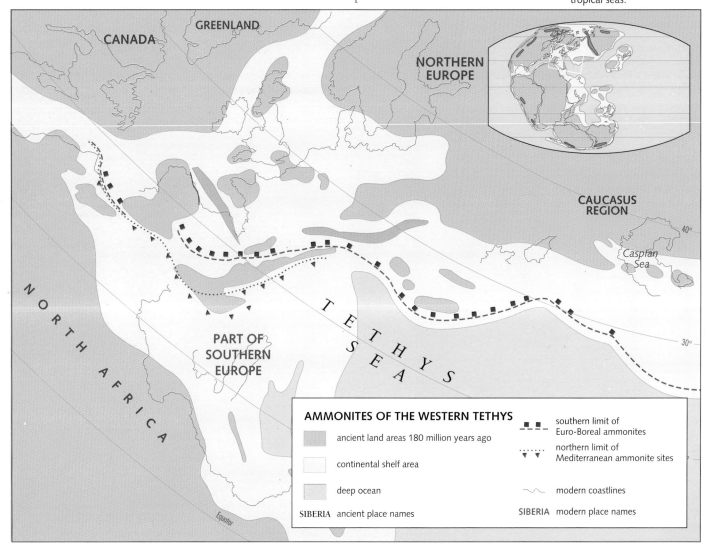

AMMONITES OF THE WESTERN TETHYS

ancient land areas 180 million years ago

continental shelf area

deep ocean

SIBERIA ancient place names

– ▪ – southern limit of Euro-Boreal ammonites

····▼··· northern limit of Mediterranean ammonite sites

～ modern coastlines

SIBERIA modern place names

The Radiation of Ammonites

Ammonites are only one part of the subclass Ammonoidea, which belongs to the cephalopod group of molluscs (the group that contains the squid and octopus). The ammonoids appeared in the Lower Devonian period (450 million years ago) and were extinct by the end of the Cretaceous period (65 million years ago). In the Mesozoic era they are a key index, or zone, fossil.

The ammonoids have a complex family history. They suffered widespread extinctions at the end of the Carboniferous and during the Permian, but diversified greatly in the Triassic. However only one family – the ceratitids – survived through to the Jurassic. One single group of the ceratitid family, the ammonites, gave rise to over 1,200 genera. The successive evolution and extinction of ammonite subgroups was so rapid during the Mesozoic, that most ammonite zones are no more than one million years in duration.

The ammonoids evolved from a similar group called the nautiloids. Only one genus from this group still survives – *Nautilus*, which lives in the Pacific, and has a similar appearance to its extinct relatives, the ammonites.

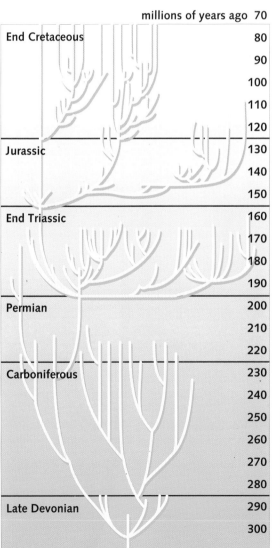

millions of years ago	70
End Cretaceous	80
	90
	100
	110
	120
Jurassic	130
	140
	150
End Triassic	160
	170
	180
	190
Permian	200
	210
	220
Carboniferous	230
	240
	250
	260
	270
	280
Late Devonian	290
	300

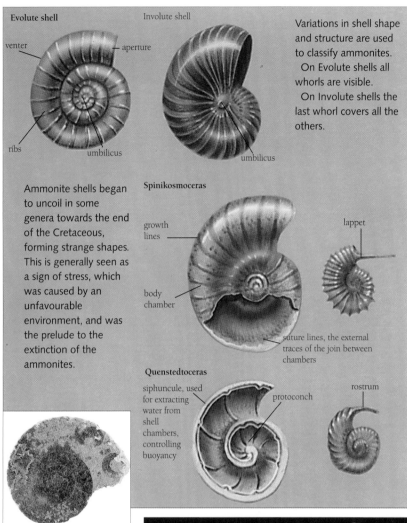

Evolute shell — venter — aperture — ribs — umbilicus — Involute shell — umbilicus

Variations in shell shape and structure are used to classify ammonites.
 On Evolute shells all whorls are visible.
 On Involute shells the last whorl covers all the others.

Ammonite shells began to uncoil in some genera towards the end of the Cretaceous, forming strange shapes. This is generally seen as a sign of stress, which was caused by an unfavourable environment, and was the prelude to the extinction of the ammonites.

Spinikosmoceras — growth lines — body chamber — lappet — suture lines, the external traces of the join between chambers

Quenstedtoceras — siphuncle, used for extracting water from shell chambers, controlling buoyancy — protoconch — rostrum

Ammonite shells
Variation in shell shape, structure and markings, make ammonites valuable date indicators. *Hudsonoceras proteum* (*above*) is used as a zone fossil in Devonian and Carboniferous strata. The crystallized shell of *Hoploscaphites nebracensis* shows (*right and below*) the internal nacre.

Nautilus
The last remaining close relative of the ammonites is the Nautilus, found in the Pacific.

Early Dinosaurs
Origins and development

Dinosaurs were the most successful vertebrate animals that ever lived on land. Members of the dinosaur family took advantage of a huge variety of ecological opportunities, diversified extensively, and dominated the Earth for a period of 150 million years. In contrast mammals, although first appearing 200 million years ago, have been ascendant for only about 50 million years, and humans emerged only in the last two to three million years.

Dinosaurs are reptiles belonging to two separate groups that evolved from the thecodont reptiles during the Triassic period. Their chief defining characteristic is their hip joint. While most reptiles have a sprawling gait with wide-apart legs, the dinosaurs developed hips that enabled them to walk upright, and to use their legs more efficiently. This development was crucial because the dinosaurs rapidly became the dominant reptile group, and the dominant land animals. Dinosaur-like reptiles also developed the ability to fly, and some took to the seas. Dinosaurs filled every available ecological niche, ranging in size from 50 cm (20 in) to 28 metres (92 ft).

Archosaur

Early dinosaurs
Archosaurs were an early reptile group, which were the probable precursors of the thecodonts. These reptiles became the dominant group in the Triassic period. Dinosaurs arose from thecodont reptiles in the mid to Late Triassic.

Triassic dinosaurs
The map (*below*) shows finds of early dinosaur fossils plotted on a map of the continents 225 million years ago. Fossil finds are inevitably grouped in areas where rocks from the period are near the surface. But the pattern of finds gives an indication of dinosaur distribution. They had already adapted to living in a wide range of latitudes.

Erythrosuchid

Euparkeria

It is not clear whether the dinosaurs arose from two different families of thecodonts, or split into two orders from a single dinosaur ancestor. By the end of the Triassic ornithiscians ('bird-hipped'),

TRIASSIC DINOSAUR SITES

ancient continents	SIBERIA	modern place names
ancient continental shelf	Theropods	
ancient mountain chains	Sauropods	
BALTICA ancient place names	Ornithiscians	
modern coastlines		

Stegosaurus

SIBERIA

KOLYMA

TETHYS

OCEAN

MANCHURIA

L A U R A S I A

GREENLAND

EUROPE

TURKEY

CENTRAL
ASIA

CHINA

SPAIN ITALY

NORTH
AMERICA

P
A
N
G
E
A

ARABIA

S E
ASIA

AFRICA

INDIA

AUSTRALIA

SOUTH
AMERICA

G O N D W A N A L A N D

ANTARCTICA

Jurassic dinosaurs

Remains of Jurassic dinosaurs have been found on every continent on Earth except Antarctica.

Most of the world's land-masses remained close to each other throughout the Mesozoic era. This made the spread of the land-based dinosaurs easier than it would have been in later or earlier times.

(left), had separated from saurischians ('reptile-hipped'). Both orders were more mobile than other reptiles because of their better hip design.

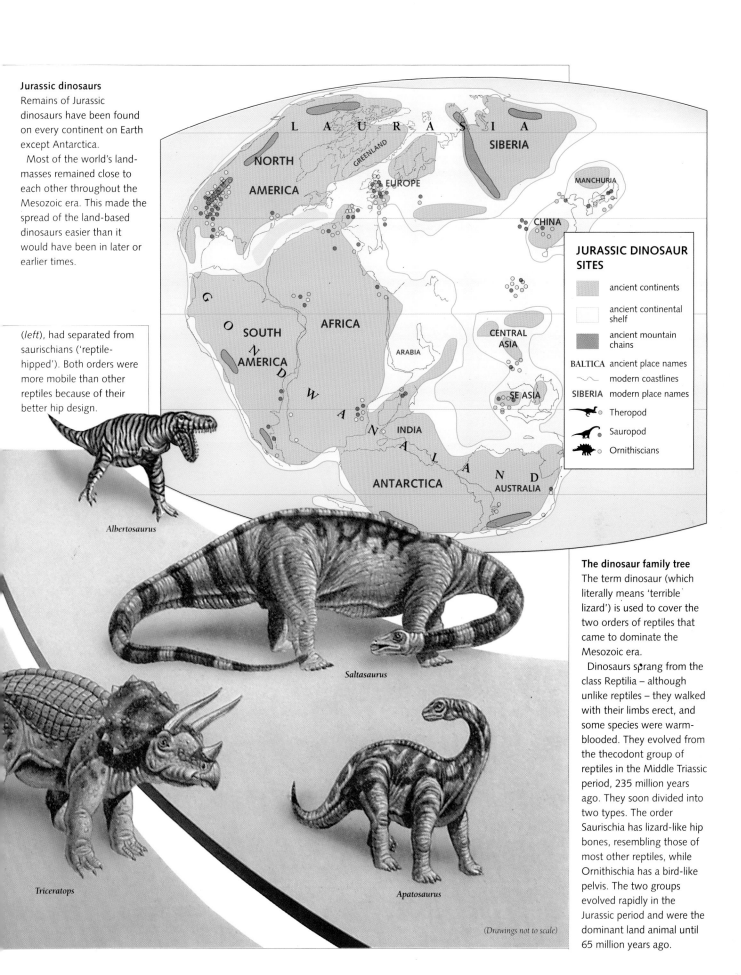

JURASSIC DINOSAUR SITES

	ancient continents
	ancient continental shelf
	ancient mountain chains
BALTICA	ancient place names
	modern coastlines
SIBERIA	modern place names
	Theropod
	Sauropod
	Ornithiscians

Albertosaurus

Saltasaurus

Triceratops

Apatosaurus

(Drawings not to scale)

The dinosaur family tree

The term dinosaur (which literally means 'terrible lizard') is used to cover the two orders of reptiles that came to dominate the Mesozoic era.

Dinosaurs sprang from the class Reptilia – although unlike reptiles – they walked with their limbs erect, and some species were warm-blooded. They evolved from the thecodont group of reptiles in the Middle Triassic period, 235 million years ago. They soon divided into two types. The order Saurischia has lizard-like hip bones, resembling those of most other reptiles, while Ornithischia has a bird-like pelvis. The two groups evolved rapidly in the Jurassic period and were the dominant land animal until 65 million years ago.

Arabian Oil

The world's largest concentration of mineral wealth

More than two-thirds of the world's known oil reserves are concentrated in the region around the Persian Gulf. A special set of geological circumstances arose in this area to produce this mineral wealth. The initial formation of oil is a fragile process. The region must be stable for a long enough period to allow the accumulation of large amounts of dead organic matter, which must then be compressed beneath further sediments. This process must be followed by fairly gentle disturbances which produce the folding and faulting that creates oil-traps.

The Arabian continental plate has remained in low latitudes – within 30° of the equator – for the last 250 million years. Oil is formed from the bodies of marine organisms that thrive in warm seas, so the region's history has allowed a great accumulation of oil deposits.

The region's oil was formed about 150 million years ago, but it has migrated from its original location. Arabia collided with Asia about 20 million years ago. The boundary between the two plates is marked by the Zagros thrust fault that runs along the north of the Persian Gulf. The impact created folds in the overlying rocks into which the oil from below migrated. The upward curves of the folds, where most of the region's oil has accumulated, are known as anticlines.

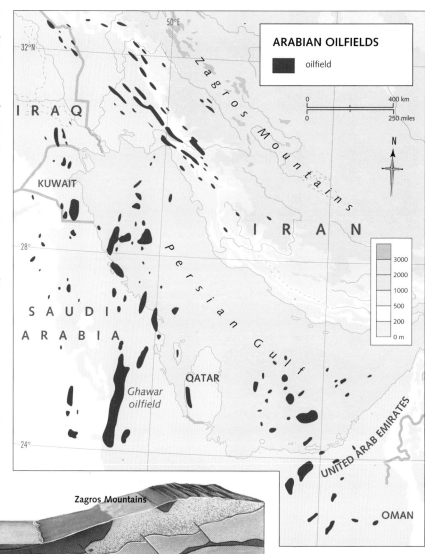

ARABIAN OILFIELDS

oilfield

0 — 400 km
0 — 250 miles

N

3000
2000
1000
500
200
0 m

IRAQ

KUWAIT

Zagros Mountains

IRAN

Persian Gulf

SAUDI ARABIA

Ghawar oilfield

QATAR

UNITED ARAB EMIRATES

OMAN

Zagros Mountains

Arabia

Persian Gulf

petroleum accumulations

salt deposits   continental basement

Paleozoic

Mesozoic

Neogene

The Oilfields of the Middle East

The region around the Persian Gulf has remained near to the equator for the past 250 million years. This, combined with a fairly stable geological history has created the conditions for the production and retention of vast amounts of oil.

The elongated oil fields that run northwest–southeast are trapped in folds created at the same time as the Zagros Mountains, 25 to 15 million years ago, though the oil comes from Mesozoic rocks, 150 million years old. Oilfields which run north–south are located in structures caused by plate movements, while the circular fields are trapped under salt structures.

Oil traps

Oil is trapped where the reservoir rocks, usually porous sandstones or limestones, are overlain by an impermeable rock such as shale, mudstone or salt. The most effective oil traps are in the domes of folded rock formations (anticlinal traps); 80 to 90 per cent of the world's oil is held in this form of trap where the light oil and gas migrates to the highest point. The vast Ghawar oilfield in Saudi Arabia is the largest known oilfield of this type.

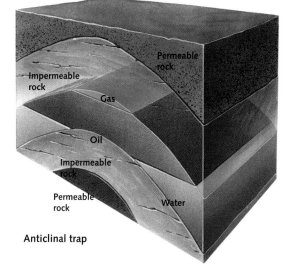

Permeable rock

Impermeable rock

Gas

Oil

Impermeable rock

Permeable rock

Water

Anticlinal trap

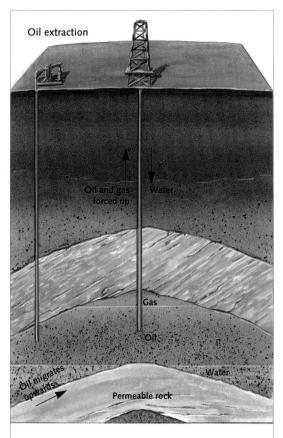

Oil extraction

Oil and gas forced up

Water

Gas

Oil

Oil migrates upwards

Permeable rock

Water

Oil Extraction

Oil reserves are found in porous rocks; non-porous strata cap the reserves and trap the oil. A pipe is drilled through the impervious layer. Water or mud is then pumped into the porous rock at high pressure. This forces the oil out, back up the pipe where it can be captured and stored. Only about one-third of the oil in the reservoir rock is recovered by these traditional methods.

Off-shore oil platform (*above*)
Many of the world's oil fields lie under continental shelf seas. This platform in the Persian Gulf is similar to those used in the North Sea, Gulf of Mexico, the South China Sea and the Arctic Ocean.
Desert pipelines (*right*)
Oil is piped direct from well-heads across Saudi Arabia to refineries and tanker jetties.

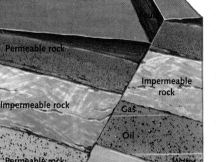

Permeable rock

Impermeable rock

Impermeable rock

Gas

Oil

Permeable rock

Water

Impermeable rock

Fault traps

Where rocks have been faulted, impermeable rocks may be thrown up against reservoir rocks, to provide an effective seal for the oil and gas they contain. The Brent Field in the North Sea is an example of this.

Some disturbance is essential for the folding and faulting which creates oil-traps to occur. But too much intense gological activity may allow the oil and gas to escape to the surface.

World Oil Reserves

There are estimated to be enough world oil reserves to last for about another 80 years. New discoveries may extend this period, but by 100 years time, oil stocks will be seriously depleted at current rates of exploitation. New research is being carried out into ways of extracting the two-thirds of oil that remains in the exploited fields.

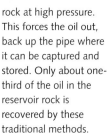

Saudi Arabia 26.0%

Other 18.7%

Kuwait 10.75%

U.S.S.R. 10.4%

Iran 9.0%

Mexico	4.9%
Iraq	4.8%
U.A. Emirates	4.6%
United States	4.1%
Libya	3.7%
Peoples's Republic of China	3.1%

The Cretaceous World

The Earth 145 to 65 million years ago

The Cretaceous period is named after the Latin word for chalk (*creta*) which is the characteristic rock of this period on the northern continents. During the Cretaceous, warm shallow seas spread over many of the continents; at the start of the period, levels were 25 metres (80 ft) above present levels. A world-wide flooding, known as the Cenomanian transgression, led to very high seas in the mid-Cretaceous – nearly 200 metres (650 ft) above present levels. This is partly explained by the fact that there was a warm overall climate and no polar glaciation in the Cretaceous period.

The Cretaceous is the last period of the Mesozoic era, which ended with a massive change in the dominant life-forms on the planet. The remains of Cretaceous dinosaurs are found on every continent; they were the dominant vertebrates of the period. By the Cretaceous, dinosaurs and other reptiles had become extremely diverse, on land, in the air and in water. Their extinction, along with many other species, at the end of the period is one of the great mysteries of the history of the Earth. Although mammals were a relatively insignificant life form during the Cretaceous they survived the extinction, and subsequently flourished.

MID CRETACEOUS 120 MILLION YEARS AGO

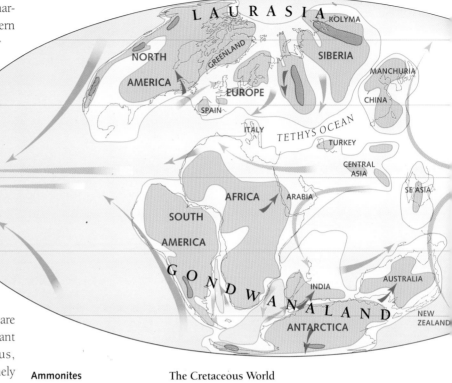

Ammonites
Ammonites (*bottom left*) are the dominant fossils of the Cretaceous period though they became extinct at the end of the period. Some individuals reached a diameter of 2 m (6 ft) Bivalves and gastropods were also widespread and are used as zone fossils where ammonites are absent.

The Cretaceous World
The break-up of the continents continued: Africa, India and parts of Asia broke away from Antarctica; South America and Africa also split apart. This produced widespread volcanic activity.

The Central Atlantic began to open as North America pulled away from Europe, while the great Tethys Ocean, which had separated Europe and Siberia from Africa and southern Asia, now began to close. The collision of Africa with Europe would lead to the Alpine Orogeny.

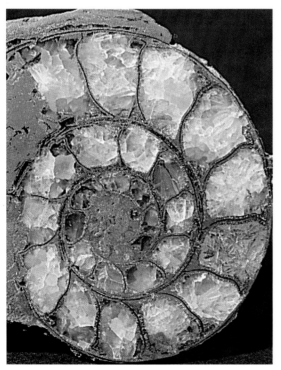

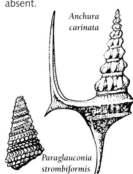

Paraglauconia strombiformis

Anchura carinata

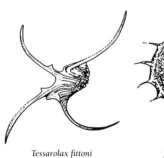

Tessarolax fittoni

Jurassiphorus fittoni

Fossil groups
Numerically, the most important fossils of the Cretaceous period are the billions of microscopic planktonic organisms preserved in the chalk formations of the northern continent. These plankton served as an important food-source for marine animals of all types. Many fossils of shelled marine animals show changes in shell shapes in the latter stages of the Cretaceous period. Long spines are common on gastropods and ammonites (*above*) – which also showed a tendency to uncurl. These changes are thought to be a response to environmental stress, probably caused by changes in sea water composition. Some groups adapted and survived these changes, but others, including the ammonites, died out.

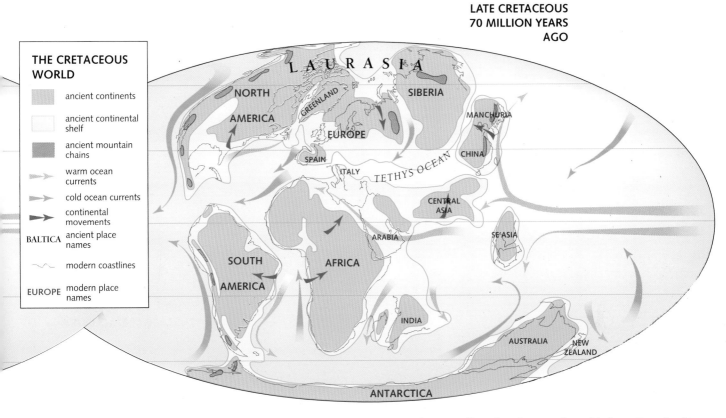

**LATE CRETACEOUS
70 MILLION YEARS
AGO**

THE CRETACEOUS
WORLD

- ancient continents
- ancient continental shelf
- ancient mountain chains
- warm ocean currents
- cold ocean currents
- continental movements

BALTICA ancient place names

modern coastlines

EUROPE modern place names

Flora and Fauna

Dinosaurs dominated the land areas of the Cretaceous. As the period progressed, groups were isolated by the movement of continents. All dinosaur groups were extinct by the end of the period. Flying reptiles, or pterosaurs, were still present, and two diving birds emerged – *Ichthyornis* and *Hesperornis*. The first snakes appeared at the end of the period.

Mammals were still a relatively insignificant life-form. Typically shrew-sized, the largest was the size of a cat. These mainly nocturnal creatures are thought to have been fairly common by the end of the Cretaceous.

Flowering plants first appeared in the Middle Cretaceous, and were the dominant land flora by the end of the period. Flora of the Late Cretaceous included figs, willows, poplars, magnolias and plane trees.

Cretaceous landscape

Dinosaurs were still dominant when flowering plants emerged and spread over the land. While dinosaurs have become extinct, flowering plants have become the dominant land plants – and the principal food source of mammals.

Chalk

Warm, tranquil waters over northern Europe

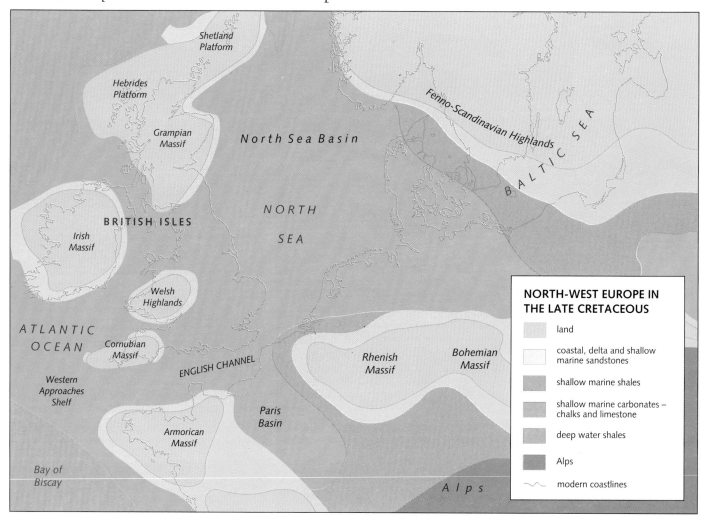

Shetland
Platform

Hebrides
Platform

Grampian
Massif

North Sea Basin

Fenno-Scandinavian Highlands

BALTIC SEA

BRITISH ISLES

NORTH

SEA

Irish
Massif

Welsh
Highlands

ATLANTIC
OCEAN

Cornubian
Massif

ENGLISH CHANNEL

Rhenish
Massif

Bohemian
Massif

Western
Approaches
Shelf

Paris
Basin

Armorican
Massif

Bay of
Biscay

A l p s

**NORTH-WEST EUROPE IN
THE LATE CRETACEOUS**

land

coastal, delta and shallow
marine sandstones

shallow marine shales

shallow marine carbonates –
chalks and limestone

deep water shales

Alps

modern coastlines

During the Cretaceous period, about 80 million years ago, sea levels rose to such an extent that almost all the continents of the Earth were under water. Shallow seas invaded the continents leaving only small areas remaining as land. The seas ebbed and flooded, and different parts of the land were under water at different times. But it seems that almost every part of the Earth was under water at some stage during the Late Cretaceous period.

For the present Earth, the result of the Cretaceous floods is seen in the massive sequences of Cretaceous chalk which are present on many continents, particularly in northern Europe. The large amounts of Cretaceous chalk tell us a great deal about the state of the Earth 80 million years ago. Chalk is a very pure, fine-grained form of limestone, which requires a particular set of conditions for its formation. By the Late Cretaceous in Europe, the period of intense erosion of the older Caledonian and Hercynian mountain ranges was over. This meant that the amount of muds and sandstones

being brought into the coastal seas by rivers was reduced. This was also a relatively tranquil period. These two factors allowed very pure carbonate to form in remarkable thicknesses

Chalk is made up of the bodies of tiny single-celled algae called coccoliths. These are only formed in clear, warm, tranquil water – usually found on continental shelf areas in the tropics. As they die, billions of microscopic coccoliths fall to the sea bed in a continuous rain to form an ooze. This process is now happening in the Caribbean. As more sediment is piled on top the ooze hardens and becomes lithified, forming the pure white rock known as chalk..

Chalk's characteristic white colour is instantly recognizable in the cliffs on either side of the English Channel. Chalk landscapes are also easy to spot. Because it soaks up water, chalk forms dry valleys and erodes into low, rounded hills.

The effect of the very high sea levels in the Cretaceous on land-based life forms was profound, leading to the extinction of many species, including groups of dinosaurs.

Chalk and Oil
When algae, or coccoliths, (*above*) are compressed by sediment into layers of chalk the tiny amount of organic material that makes up the soft bodies of the coccoliths is squeezed out. This fluid, rich in carbon, is the basis of crude oil.

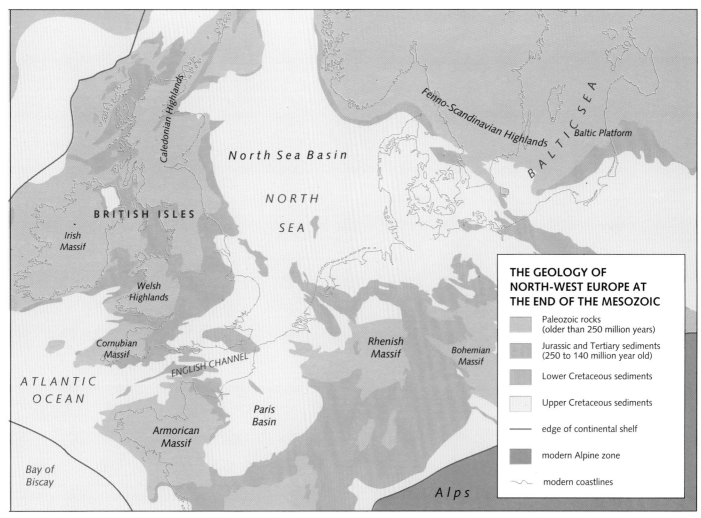

THE GEOLOGY OF
NORTH-WEST EUROPE AT
THE END OF THE MESOZOIC

- Paleozoic rocks
  (older than 250 million years)
- Jurassic and Tertiary sediments
  (250 to 140 million year old)
- Lower Cretaceous sediments
- Upper Cretaceous sediments
- edge of continental shelf
- modern Alpine zone
- modern coastlines

The geology of north-west Europe

The upland areas of northern Europe had been created during the Caledonian and Hercynian Orogenies by collisions with North America and with Africa. During the Mesozoic era the lowlands and basins between these hills were repeatedly filled with eroded sediment, flooded by shallow seas or drained and dried by the tropical sun. The Late Cretaceous period, 80 million years ago, was a time of huge flooding of the Earth's continents, with global sea levels about 200 metres (650 ft) above their present levels. Large parts of the continent of Europe were covered by shallow sea (*map above left*). The old uplands stayed above sea level and were surrounded by beaches and river deltas, depositing sand and mud. Warm, tropical seas covered the lowlands in conditions that were perfect for the formation of very pure carbonates – i.e. chalk.

The distribution of Mesozoic and Paleozoic rocks in the same area (*map above*) shows that all the Cretacous sediments were created in these shallow seas, explaining why virtually all Cretaceous rocks in northern Europe are chalk. The great chalk deposits of the North Sea basin are the source of the oil and gas discovered in abundance since the 1970s. New oilfields are still being discovered in the northern North Sea.

The White Cliffs of Dover
During the Cretaceous period the region around the British Isles became relatively stable and tranquil, covered by tropical seas. While an area of higher land known as 'London Island' may have remained above sea level together with the lands of north-west Britain, the rest of the region was blanketed in a thick layer of chalk – reaching over 500 m (1,650 ft) in places. Cretaceous chalk formations lie on either side of the English Channel, and form familiar scenery such as the White Cliffs of Dover (*above*) and the Seven Sisters. At Beachy Head, the cliffs reach a height of nearly 200 m (650 ft).

Chalk downland, Amberley, Sussex
Chalk is slightly soluble in water, and extremely porous. These two properties led to the formation of characteristic landscapes in chalk areas. The lack of trees on chalk hills was a reason why these were regions of early human settlement in Europe.

The Arctic Basin and Verkhoyansk Mountains
Ocean floor and oil-rich shelf seas

The formation of the Arctic Basin has had an important effect on the Earth. The changing positions of the continents over the past 500 million years has generally allowed ocean currents to flow freely into the polar regions. However, in the last 10 million years the Arctic has become encircled by continental blocks which have effectively cut off the supply of warm water. This has helped to make the Earth gradually colder with ice caps now a permanent feature of the north and south polar regions. The process of isolating the Arctic was accelerated in the Cretaceous period by the disturbances which led to the formation of the Verkhóyansk Mountains.

Baltica, Siberia, Kazakhstan and Kolyma were originally formed as separate small continents. They remained separate, but closely related, throughout the Paleozoic era. Baltica became joined to Laurentia during the long Caledonian Orogeny. Siberia and Kazakhstan became joined and then collided with Baltica at the end of the Paleozoic era 250 million years ago, forming the Ural Mounatins. Kolyma, which now sits at the far north-eastern corner of Asia, travelled across the North Pole and collided with Siberia in the Cretaceous period. The resulting disturbance threw up the Verkhoyansk Mountains, and formed the Arctic Basin.

THE ARCTIC BASIN

modern coastline

2000m below sea level

sedimentary basin

The Arctic Basin

The present Arctic Basin is one of the great oceans of the world. It was formed in two separate tectonic processes. During the first episode, the small continent of Kolyma was pushed across the North Polar region as a zone of sea floor spreading opened up a new ocean. The second episode came 70 million years later, as the northern North Atlantic opened. North America was pushed to the west and Eurasia to the east, until they met at the Bering Straits. This opened up a basin between the Lomonosov ridge and Siberia, and effectively encircled the Actic basin with land.

The continental movements around the Arctic have created a large number of oil traps. Oil and gas are formed in sedimentary basins by the accumulation of dead marine organisms, and their subsequent conversion into hydrocarbons. Both crude oil and gas are extremely light and tend to migrate upwards through rocks, particularly as pressure is applied from above. They are only prevented from escaping onto the surface if there are non-porous overlying rocks (*page 104*). There are hundreds of old sedimentary basins on the continental margins around the Arctic Basin, most holding reserves of oil and gas. The overall reserves of the Arctic region are massive, but much of the region remains unexplored, and the difficulties of working in permafrost have delayed the exploitation of the region's reserves.

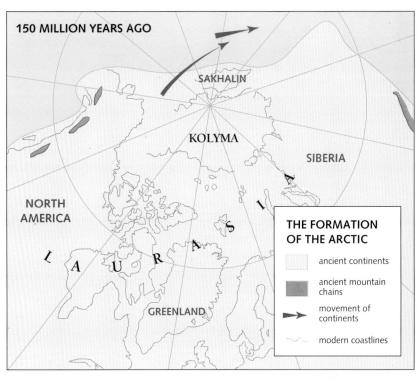

Arctic oil installation
Exploitation of Alaskan oil began in earnest in the 1970s. This was the first part of the Arctic to be opened up, as the resources of the American oil industry were available to fund exploration and the development of necesary new technologies. Transportation is a huge problem in regions where there are no ice-free ports and no roads or railway lines. This applies to most of the Siberian and Canadian Arctic. The reserves here are thought to be immense – and pipelines are being built to link drilling platforms to refineries and centres of population where oil will be used.

The formation of the Arctic
Europe and North America were still joined 150 million years ago (*above*). By the Jurassic period Siberia was joined to Laurasia. The small continent of Kolyma lay on the northern margin of this massive continent with the island of Sakhalin lying offshore.
120 million years ago (*below*) Kolyma and Sakhalin moved across the North Pole and collided with the north-east edge of Siberia, forming the Verkhoyansk Mountains.

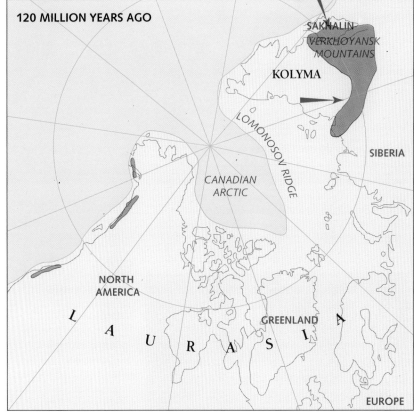

Flowering Plants
Domination of the plant kingdom

During the early Cretaceous period, a new type of plant evolved which came to dominate the landscape of the Earth. The flowering plants, or angiosperms, diversified and adapted rapidly, so that by the end of the Cretaceous they occupied a wide variety of habitats. Of the approximately 300,000 species of green plant that exist, over 250,000 are angiosperms. Flowering plants supply mammals with almost all their food, so it is no coincidence that the great diversification of mammals, including the appearance of human beings, occurred in parallel with the development of flowering plants.

These plants in return depend on all sorts of animals for their fertilization and the dispersal of their seeds. The process of pollination, fruit-bearing and seed dispersal is one of the most striking examples of how different life-forms have evolved to depend on one another. The evolutionary success of angiosperms has been helped by the colours of their flowers. Different colours attract different birds or insects. This helps to retain the genetic integrity of the species by restricting hybridization, and also enables genetic information, for example, advantageous mutations, to be spread very quickly – one bee may fertilize many plants with the same pollen.

THE SPREAD OF FLOWERING PLANTS

- ancient continents
- ancient continental shelf
- spread of flowering plants
- area of origin of flowering plants
- distribution of Nothofagus in Gondwana

The spread of flowering plants

This map shows a possible mechanism for the rapid spread of angiosperms around the world. In the early Cretaceous period, 140 million years ago, almost all the continents were still closely linked, though beginning to drift apart. Early angiosprems from the region of west Gondwanaland could have spread over land to almost all the continents. Flowering plants have in any case developed highly sophisticated methods of seed dispersal, enabling them to travel across water and other inhospitable enviroments.

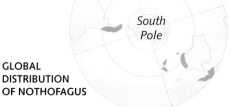

GLOBAL DISTRIBUTION OF PROTEACEAE

Proteaceae

This diverse family of angiosperms is found on all the southern continents and as far north as Mexico. Its distribution across the southern hemisphere is hard to explain without reference to continental drift and an ancient southern supercontinent which linked all these land masses.

GLOBAL DISTRIBUTION OF NOTHOFAGUS

Nothofagus

Fossilized pollen from the *Nothofagus* (Antarctic beeches) appears in 70 million year-old rocks. Seeds are dispersed by wind, but must travel over land since they are killed by salt water. There are no records of *Nothofagus* in Africa – it is likely to have arisen in South America after the split from Africa, and migrated across Antarctica to Australasia.

EPACIDACEAE

GOODENIACEAE

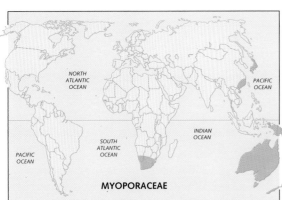

MYOPORACEAE

STYLIDIACEAE

Floral Realms

The flowering plants of the world are divided geographically into floral realms. But the realms of the present world are also a reflection of the Earth's past geography, when the continents were arranged differently. The Cape province of South Africa has more plant groups in common with India, the Antarctic and Australia than it does with the plants of the rest of Africa. The boreal floral realm reaches across all the northern continents, which were connected 100 million years ago.

The four maps show the modern distributions of, from the top, the Epacidaceae, Goodeniaceae, Myoporaceae and Stylidiaceae families of angiosperms. All have distributions which must have puzzled pioneering botanists, but which make sense when plotted on a reconstruction of Gondwanaland. These families were among the first angiosperms to develop – becoming established over a wide range of Gondwanaland, before the super-continent broke up into its constituent parts.

Late Cretaceous Forest

Woodlands of 70 million years ago resembled those of the present with deciduous trees, including oak, willow and magnolia. Grasses (also a flowering plant family) did not develop until the mid-Tertiary period.

Cretaceous Dinosaurs

Dominance, isolation and decline

At the beginning of the Cretaceous period 140 million years ago, the dinosaurs were the dominant animals on the Earth. They developed to fill a wide range of ecological opportunities and became superbly adapted to life in the Mesozoic era. Their close relations, the pterosaurs, filled the skies, and other reptiles like the plesiosaurs and mosasaurs colonized the waters of the Earth.

New dinosaur forms appeared during the Cretaceous, including toothless bird-like creatures. New plant-eaters developed protection from the large predators. These new horned and armoured herbivores, together with new duck-billed types replaced the huge long-necked sauropods. Among the meat-eaters, the tyrannosaurids were the heaviest known carnivores. By the Late Cretaceous dinosaurs were living alongside modern-looking reptiles and amphibians – for example, frogs, turtles, snakes and crocodiles.

By the time of their disappearance at the end of the Cretaceous, several dinosaur groups had become warm-blooded, allowing them to live in colder climates. It seems likely that dinosaurs continued to adapt right up to their total extinction 65 million years ago.

(Right) A skull of *Protoceratops*. This 31cm (12 in)-long skull is from a two-thirds grown juvenile. *Protoceratops* was an ornithischian, with horns and a neck frill.
(Far right) A skull of *Diplodocus*. An immense sauropod growing to 27m (89 ft). Remains have been found in late Jurassic rocks in the United States.

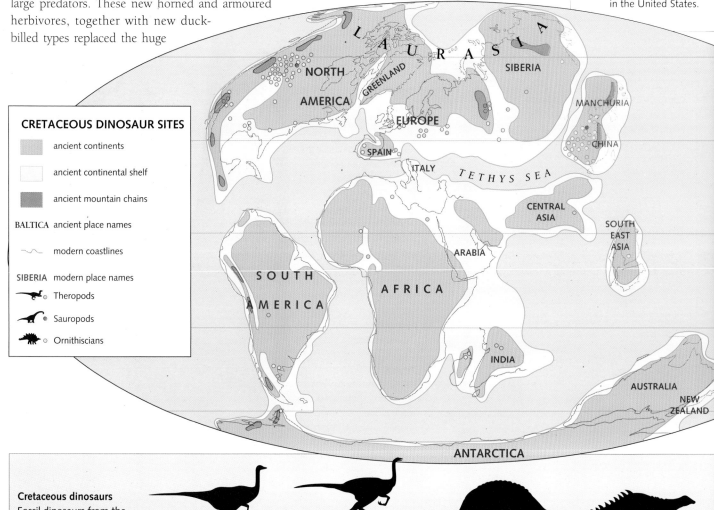

CRETACEOUS DINOSAUR SITES

- ancient continents
- ancient continental shelf
- ancient mountain chains

BALTICA ancient place names

~~~ modern coastlines

SIBERIA modern place names

- Theropods
- Sauropods
- Ornithiscians

Cretaceous dinosaurs
Fossil dinosaurs from the Cretaceous period are found on all continents. The most fertile places for fossil finds are in the semi-arid regions of North America and Asia, where fossil bones are well preserved, but the rock is eroded sufficiently to expose them.

Ornithomimus
Medium-sized Ornithischian, 3 to 4m (10ft), including a 2m (6ft) tail. The first remains were found in 1889 in Colorado.

Gallimimus
At 4m (13ft) long, this was probably the largest ornithomimosaur. Its hands may have been used for scraping at the soil. It was discovered in China.

Nodosaurus
This dinosaur was armoured with small bony knobs, and had powerful limbs to support its body weight. Skeleton finds are from western North America.

Acanthopholis
This 5.5m (18ft)-long animal was protected by rows of oval bony plates and sharp spikes, which lined its back. Its remains were first found in Kent in 1864.

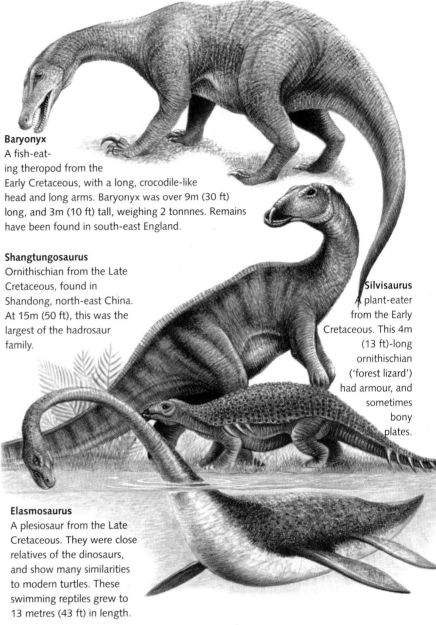

Baryonyx
A fish-eat-
ing theropod from the
Early Cretaceous, with a long, crocodile-like
head and long arms. Baryonyx was over 9m (30 ft)
long, and 3m (10 ft) tall, weighing 2 tonnnes. Remains
have been found in south-east England.

Shangtungosaurus
Ornithischian from the Late
Cretaceous, found in
Shandong, north-east China.
At 15m (50 ft), this was the
largest of the hadrosaur
family.

Silvisaurus
A plant-eater
from the Early
Cretaceous. This 4m
(13 ft)-long
ornithischian
('forest lizard')
had armour, and
sometimes
bony
plates.

The Cretaceous World
The map shows locations of significant Cretaceous dinosaur fossil finds, plotted onto a reconstruction of the world's geography 85 million years ago. Comparing this with the map of Jurassic dinosaur sites on page 103, the range of dinosaurs seems to have shrunk – this may be because Cretaceous rocks are less common in certain areas of the world, and so the fossils of dinosaurs have not been so widely preserved.

The break-up of the super-continent of Pangea continued through the Cretaceous period. This meant that the continents became separated and effectively prevented the migration of land animals . Groups were isolated on particular continents and developed different characteristics from other groups. There were other barriers to migration: 80 million years ago the eastern edge of North America consisted of high mountains which helped to cut off the large number of dinosaurs living on the plains to the west.

Elasmosaurus
A plesiosaur from the Late
Cretaceous. They were close
relatives of the dinosaurs,
and show many similarities
to modern turtles. These
swimming reptiles grew to
13 metres (43 ft) in length.

Brachylophosaurus
A duck-billed dinosaur; the first skull was found in Alberta, Canada in 1936. It was crested – possibly as a kind of identification signal for other dinosaurs.

Parasaurolophus
Remains of this duck-billed dinosaur have been found in western North America. A distinctive tubular crest, up to 1.8m (6ft) long, curved back from its snout.

Spinosaurus
This meat-eater was up to 12m (40ft) long. The spines on its back – possibly used to control body temperature – were 2 m high. It was found in North Africa.

Lambeosaurus
This crested duck-billed dinosaur reached a height of 15m (50ft). Crests in males were larger than in females. Remains have been found in western Canada.

Ankylosaurus
This dinosaur was over 10m (33ft) long, and pro-tected by spines and bone plates. It had a club-shaped tail which could be used against attackers.

South America splits from Africa
The opening of the South Atlantic

South America and Africa were joined for most of the Earth's history. They were part of the original continent of Gondwanaland formed during the Precambrian era at least 600 million years ago, and remained together as all the continents formed the supercontinent of Pangea. Early in the Mesozoic era, around 200 million years ago, changes in convection currents in the Earth's mantle caused the dismemberment of Pangea – Gondwanaland also broke up as part of this process.

The geographical fit between the eastern coastline of South America and the west coast of Africa had encouraged some geologists at the start of the century to investigate the possibility of continental movement. But it was only with the development of techniques for measuring magnetism in rocks that their original speculations were confirmed.

The opening of the South Atlantic began in the Cretaceous period about 120 million years ago, as South America began to pivot away from a point in what is now the western Sahara. By 85 million years ago the ocean basin of the Central Atlantic extended south all the way to Antarctica. The two continents have continued to move apart: South America is still moving to the west – causing continuing volcanism in the Andes. Africa collided with Europe during the Alpine Orogeny and now seems likely to be spilt by the opening of a new ocean along the line of the Red Sea and the Great Rift Valley.

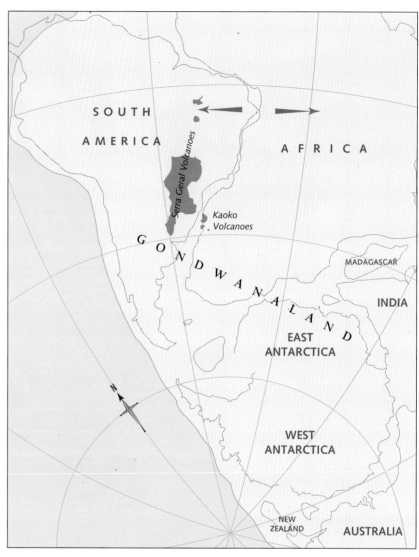

Sugar Loaf Mountain
The beautiful shapes of the mountains of Rio de Janeiro harbour (*left*) are eroded granite. Similar granites are found on the west coast of Africa – they were probably all formed in the early Paleozoic in the same orogenic episode.

The Serra Geral Mountains of southern Brazil are made of basalt which erupted during the first phase of the split between South America and Africa. Their equivalent on the African side are the Kaoka volcanoes of Namibia.

120 millon years ago
The southern supercontinent of Gondwanaland was substantially intact. South America was attached to Africa which in turn was still joined to Madagascar, India, Australia and Antarctica. Volcanic activity in Koaka, in present day Namibia, as well as in the Serra Geral in south-eastern Brazil indicates the beginnings of a split between South America and Africa.

GONDWANA 120–85 MILLION YEARS AGO

ancient continents

volcanoes

SIBERIA   ancient place names

modern coastlines

INDIA   modern place names

continental movements

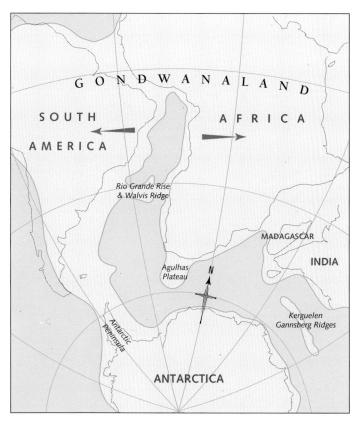

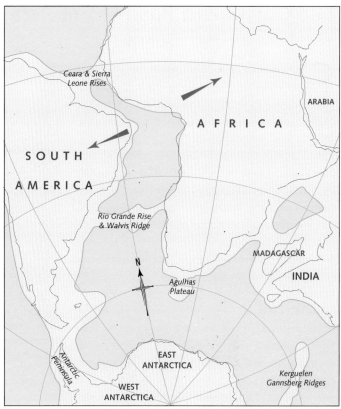

100 million years ago

New convection currents in the Earth's mantle were splitting Gondwanaland into its constituent continents. Africa, India and Madagascar formed one continental block, while South America and Antarctica were joined via the Antarctic Peninsula. The Rio Grande rise and the Walvis Ridge formed a land bridge between Africa and South America at this time.

This also cut off the movement of marine animals from the south to the north and vice versa.

Slightly later than this, as sea levels rose, a shallow sea was formed across Africa from the Atlantic to the Tethys Sea north-east of Africa. This existed for a short time around 92 million years ago. Some marine animals migrated between the two oceans causing a mixing of faunas.

Mesosaurus

This reptile (*below*) lived about 100 million years ago, and fossil remains have been found in west Africa and north-east Brazil. *Mesosaurus* was a freshwater dweller, and could not have swum across the salt waters separating these two continents. The answer is that at that time the continents were joined – compelling evidence for Continental Drift.

85 millon years ago

Though the continental shelves of South America and Africa were still linked via the Rio Grande Rise and Walvis Ridge, the north and south parts of the South

Atlantic were linked by shallow seawater. From now on the development of the South Atlantic followed the initial pattern, as both basins continued to widen and deepen.

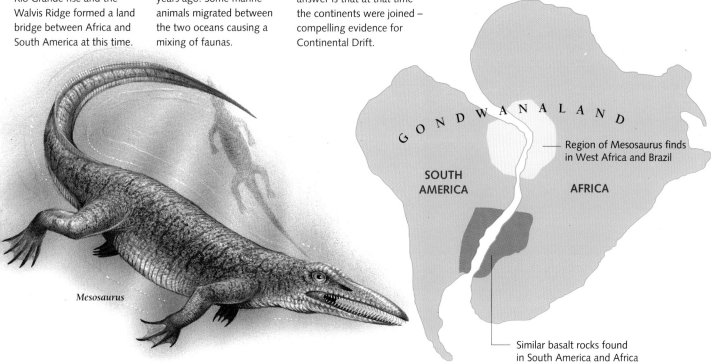

Mesosaurus

Region of Mesosaurus finds in West Africa and Brazil

Similar basalt rocks found in South America and Africa

The Caribbean Basin

A sea between two continents

For the last 110 million years North and South America have been involved in a series of movements which formed the Caribbean basin, Central America and, about 3 million years ago, the land bridge between the continents.

The Atlantic seafloor, which is spreading west, meets the eastward push of the Pacific in this region, and the separate Caribbean plate sits between. At its western edge the Pacific plate is being subducted, causing volcanic activity and earthquakes in Central America. On its eastern margin the destruction of the Atlantic seafloor has created the volcanic island arc of the Lesser Antilles (Leeward and Windward Islands).

For most of their history the continents of North and South America have remained separate. North America was the main constituent of the ancient continent of Laurentia, which was closely associated with northern Europe for the hundreds of millions of years of the Paleozoic era. South America, together with Africa, India, Australia and Antarctica was part of the continent of Gondwanaland. The regrouping of the continents into their present configuration has brought about conjunctions unprecedented in the Earth's history.

Faunal Exchanges

The joining of North and South America enabled land-based animals to migrate across the new land bridge. The two continents had been separated for 150 million years. During that time mammal groups had evolved quite independently in each continent. South American mammals were a mixture of marsupials and placentals which were unique to that continent, while those in North America were placentals. Placental mammals are generally more effective at reproducing and so some North American species have replaced their marsupial equivalents in the South. In general there have been more migrations from north to south than vice versa.

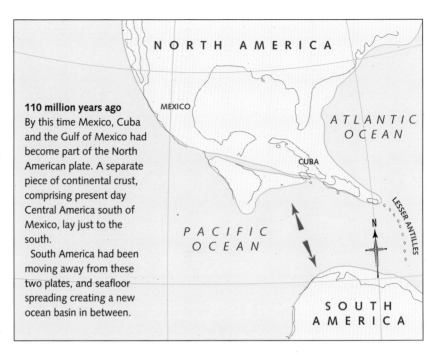

110 million years ago
By this time Mexico, Cuba and the Gulf of Mexico had become part of the North American plate. A separate piece of continental crust, comprising present day Central America south of Mexico, lay just to the south.

South America had been moving away from these two plates, and seafloor spreading creating a new ocean basin in between.

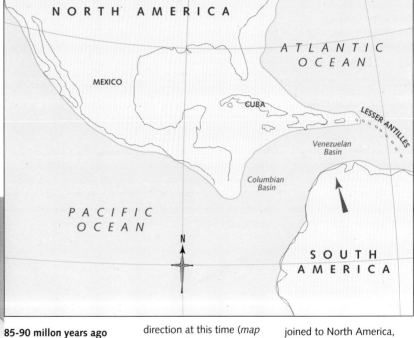

85-90 millon years ago
The sea floor spreading in the Atlantic, to the west of the Caribbean, changed direction at this time (*map above*). The spreading of the South Atlantic (between South America and Africa) became linked to the Central Atlantic (between North America and Europe/Africa). The effect was to push the three plates closer together. Central America became joined to North America, while South America moved north, compressing the newly formed ocean basin. About 1,000 kilometres (650 miles) of ocean basin was lost in subduction zones along the north and south margins of what is now the Venezuelan and Colombian basins. The Atlantic seafloor spreading also led to subduction along an arc which created the volcanic islands of the Lesser Antilles.

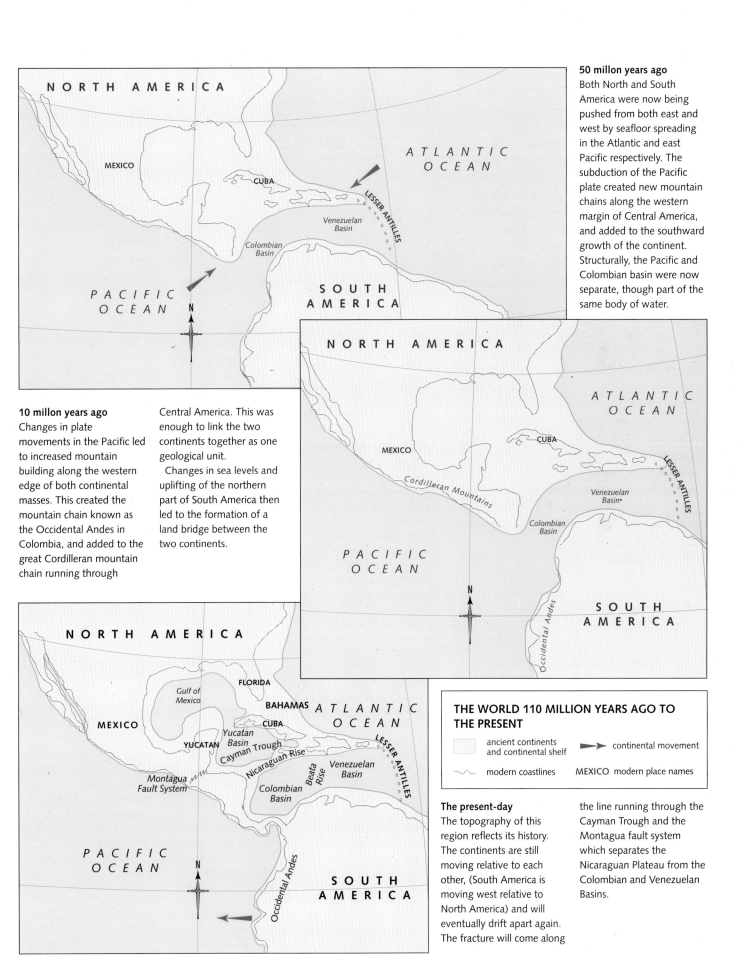

50 millon years ago
Both North and South America were now being pushed from both east and west by seafloor spreading in the Atlantic and east Pacific respectively. The subduction of the Pacific plate created new mountain chains along the western margin of Central America, and added to the southward growth of the continent. Structurally, the Pacific and Colombian basin were now separate, though part of the same body of water.

10 millon years ago
Changes in plate movements in the Pacific led to increased mountain building along the western edge of both continental masses. This created the mountain chain known as the Occidental Andes in Colombia, and added to the great Cordilleran mountain chain running through Central America. This was enough to link the two continents together as one geological unit.

Changes in sea levels and uplifting of the northern part of South America then led to the formation of a land bridge between the two continents.

THE WORLD 110 MILLION YEARS AGO TO THE PRESENT

ancient continents and continental shelf	continental movement
modern coastlines	MEXICO modern place names

The present-day
The topography of this region reflects its history. The continents are still moving relative to each other, (South America is moving west relative to North America) and will eventually drift apart again. The fracture will come along the line running through the Cayman Trough and the Montagua fault system which separates the Nicaraguan Plateau from the Colombian and Venezuelan Basins.

Mass Extinction of Mesozoic Life

From dinosaurs to mammals

The end of the Cretaceous is also the end of the Mesozoic ('middle life') era. This was a time of huge significance in the Earth's history, as the mass extinction of large numbers of dominant animals led to a change in the nature of life on Earth. The decline of the land reptiles, particularly the extinction of the dinosaurs, allowing the extraordinary development of the most dominant animal group of the last 65 million years – the mammals.

The most important fossil group of the Mesozoic era, the ammonites, also became extinct at the end of the Cretaceous period – showing that whatever changes were brought about, the extinctions affected marine life and land-based animals. Fish groups, however, remained almost unaffected, and many reptiles that were close relatives of the dinosaurs survived to thrive in the Tertiary period.

Arguments about the cause or causes of the demise of so many animal groups were transformed in 1980 by the discovery of traces of the element iridium in rocks at the boundary between the Cretaceous and Tertiary periods. These discoveries were made by the Alvarezes, a father and son team from the University of California. Since the initial report, the iridium-enriched layer has been found to cover the entire Earth. The so-called Alvarez hypothesis is that the iridium is evidence of a large asteroid impact, which would have affected the Earth's climate to such a degree that mass extinctions resulted. But other scientists have since argued that the extinctions were gradual, not catastrophic, and were probably caused by a gradual change in climate, punctuated by sudden episodes of rapid change. Drilling of deep sea sediments has revealed fluctuations and an overall decline in populations of marine plankton – an essential climate indicator and a key element in the food chain – throughout the Late Cretaceous. We are used to thinking of this most dramatic of mass extinctions as a single event. But different groups of animals were affected in different ways. Some declined gradually throughout the Cretaceous, while others flourished until near the end, and then became extinct. Still others adapted successfully enough to survive and flourish into the following Tertiary period.

The Cretaceous was a period of widespread change. Great variations in sea levels saw flooded continental margins giving way to exposed delta plains. The climate cooled, and there were significant plate movements, with major mountain-building episodes and intensive volcanic activity.

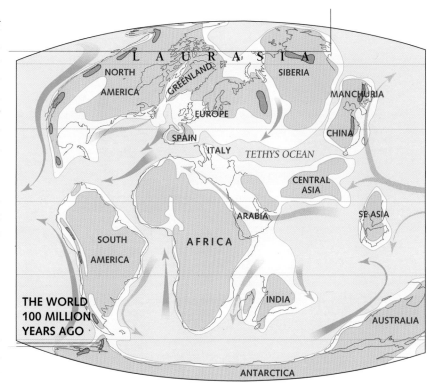

THE WORLD 100 MILLION YEARS AGO

Ocean currents and climate change

The world's climate is likely to have been significantly affected by the joining of the continents of North and South America (*see maps*). Although the land bridge between them only formed recently, the two Americas were part of one continental block by the end of the Cretaceous, causing an enormous shift in the ocean currents which are the driving force of the Earth's climate.

The effects of the changing physical geography of the Earth on its climate has been understood in general terms for some time. But it has only recently been suggested that alterations in the configuration of the continents can induce a sudden and profound change in global climatic conditions. Any such change would, in turn, affect the life of the planet. One of the peculiarities of the mass extinctions at the end of the Mesozoic is the range of life forms that were affected.

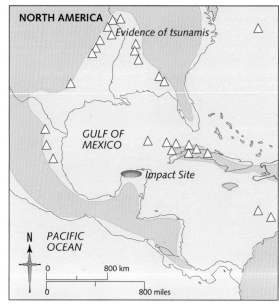

Asteroids and iridium

In 1980 it was discovered that a thin layer of clay, lying at the top of the Cretaceous levels, was enriched in the element iridium. This enrichment is worldwide, and coincides with the sudden extinction of formaniferal plankton. The presence of iridium cannot be explained by normal surface processes, and so the impact of a large asteroid with the Earth has been suggested. No collision site has been confirmed, but heating of rocks and evidence of tidal waves, make the Mexican site of Chicxhluk a strong probability.

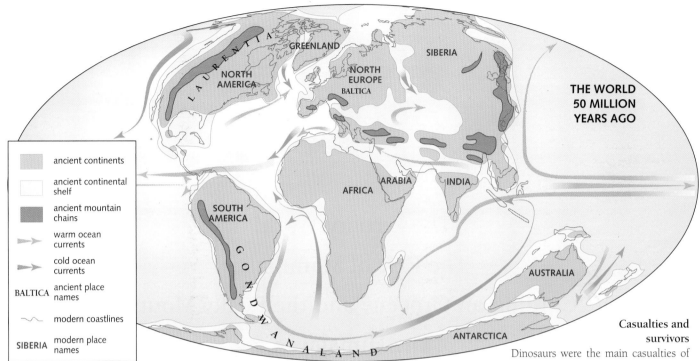

ancient continents

ancient continental shelf

ancient mountain chains

warm ocean currents

cold ocean currents

BALTICA ancient place names

modern coastlines

SIBERIA modern place names

THE WORLD
50 MILLION
YEARS AGO

The Deccan traps

At the end of the Cretaceous, as Gondwanaland and Laurasia continued to fragment, there was a vast eruption of basalt onto the surface of the Indian continent. The basalt is still there, covering a large area known as the Deccan plateau. This cataclysmic eruption would probably have had a severe impact on the Earth's climate. It is also possible that this could have been resposible for the iridium-enriched layer in Cretaceous rocks. Eruptions of basalt often coincide with the rifting apart of continents, and are caused by magma erupting directly from the mantle through newly opened fissures in the crust.

Casualties and survivors

Dinosaurs were the main casualties of the Mesozoic extinctions: only one group of dinosaurs, the triceratopsids, is known in very late Cretaceous deposits. The giant marine reptiles, the ichthyosaurs and plesiosaurs were in marked decline in the Cretaceous after strong development in the Jurassic. And the mosasaurs (marine lizards) evolved rapidly in the early Cretaceous only to become extinct by the end. Lizards and turtles, however, were able to adapt to the new Tertiary environment and continued to flourish. The small mammals of the early Cretaceous diversified during the period, producing the first insectivores and marsupials. These new types underwent rapid evolutionary radiation in the Tertiary. Birds, which were rapidly becoming better adapted to flight, were unaffected.

Struthiosaurus

Nodosaurus

Ankylosaurus

Part Six
The Tertiary Era

The Tertiary World
The Earth 65 to 2 million years ago

The Early Tertiary period saw the continents moving closer to their present positions, and the geography of the Earth becoming recognizable to us, although catastrophic geological events over the next 50 million years were still to change large areas of the Earth's surface.

During the Tertiary period the North and South Atlantic were both expanding, pushing Europe away from North America, and Africa away from South America. By 25 million years ago the Indian continent was making contact with the southern edge of the main Asian plate. Asia was yet to be formed into one landmass: the collision of India with Asia created the Himalayas.

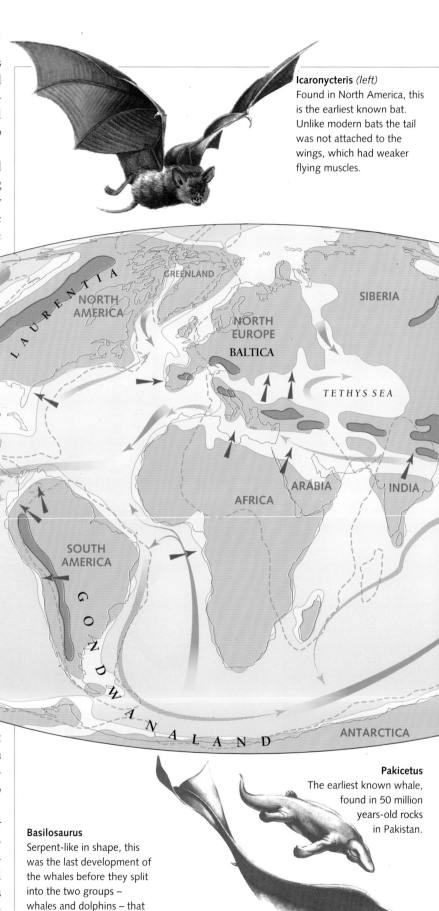

Icaronycteris (left)
Found in North America, this is the earliest known bat. Unlike modern bats the tail was not attached to the wings, which had weaker flying muscles.

THE TERTIARY WORLD
The world 50 million years ago:

- ancient continents
- ancient continental shelf
- ancient mountain chains
- warm ocean currents
- cold ocean currents
- **BALTICA** ancient place names
- modern coastlines
- coastline 25 million years ago
- **SIBERIA** modern place names
- continental movement

Further north and west the African continent was approaching southern Europe. This was to lead to the large-scale mountain-building of the Alpine Orogeny. The movement of Africa also created a depression between Africa and southern Europe by buckling the crust downwards. This depression later filled with water to become the Mediterranean Sea.

South and North America were getting closer to each other, but a major warm current still circulated between them and round the equator. Ocean currents brought warm water to the Antarctic and brought cold water away. This kept the polar regions comparatively warm. The Earth in the Tertiary was on average several degrees warmer than today.

Basilosaurus
Serpent-like in shape, this was the last development of the whales before they split into the two groups – whales and dolphins – that we see today.

Pakicetus
The earliest known whale, found in 50 million years-old rocks in Pakistan.

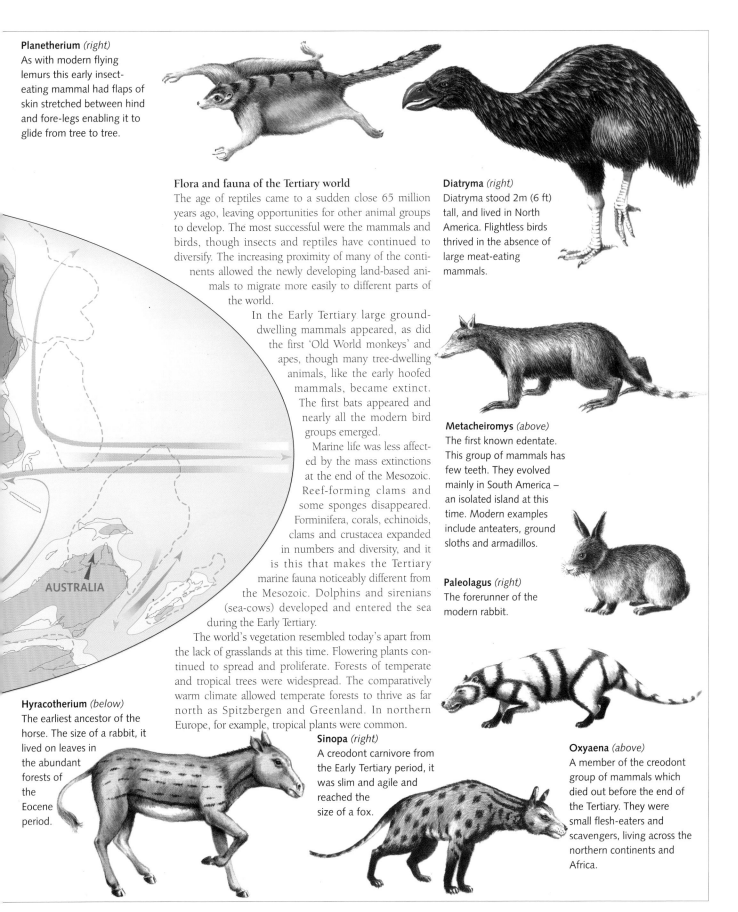

Planetherium (right)
As with modern flying lemurs this early insect-eating mammal had flaps of skin stretched between hind and fore-legs enabling it to glide from tree to tree.

Diatryma (right)
Diatryma stood 2m (6 ft) tall, and lived in North America. Flightless birds thrived in the absence of large meat-eating mammals.

Flora and fauna of the Tertiary world

The age of reptiles came to a sudden close 65 million years ago, leaving opportunities for other animal groups to develop. The most successful were the mammals and birds, though insects and reptiles have continued to diversify. The increasing proximity of many of the continents allowed the newly developing land-based animals to migrate more easily to different parts of the world.

In the Early Tertiary large ground-dwelling mammals appeared, as did the first 'Old World monkeys' and apes, though many tree-dwelling animals, like the early hoofed mammals, became extinct. The first bats appeared and nearly all the modern bird groups emerged.

Marine life was less affected by the mass extinctions at the end of the Mesozoic. Reef-forming clams and some sponges disappeared. Forminifera, corals, echinoids, clams and crustacea expanded in numbers and diversity, and it is this that makes the Tertiary marine fauna noticeably different from the Mesozoic. Dolphins and sirenians (sea-cows) developed and entered the sea during the Early Tertiary.

The world's vegetation resembled today's apart from the lack of grasslands at this time. Flowering plants continued to spread and proliferate. Forests of temperate and tropical trees were widespread. The comparatively warm climate allowed temperate forests to thrive as far north as Spitzbergen and Greenland. In northern Europe, for example, tropical plants were common.

AUSTRALIA

Metacheiromys (above)
The first known edentate. This group of mammals has few teeth. They evolved mainly in South America – an isolated island at this time. Modern examples include anteaters, ground sloths and armadillos.

Paleolagus (right)
The forerunner of the modern rabbit.

Hyracotherium (below)
The earliest ancestor of the horse. The size of a rabbit, it lived on leaves in the abundant forests of the Eocene period.

Sinopa (right)
A creodont carnivore from the Early Tertiary period, it was slim and agile and reached the size of a fox.

Oxyaena (above)
A member of the creodont group of mammals which died out before the end of the Tertiary. They were small flesh-eaters and scavengers, living across the northern continents and Africa.

The Rise of Mammals

Development and spread of the dominant land animals

Mammals are now the dominant vertebrate animals on the Earth. But for the first 150 million years of their existence they were mostly tiny creatures, living in the shadow of the great reptiles. A particular group of reptiles, the *therapsids*, had begun to develop mammal-like characteristics during the Permian period (about 270 million years ago). The therapsids diversified and became the dominant reptile group throughout the Triassic. Some species grew hair in place of scales, and became warm-blooded. These were the ancestors of the mammals.

The first true mammals appeared in the late Triassic, about 220 million years ago. The earliest known fossil is of a shrew-like creature called Megazostrodon, which has been found in Triassic rocks. Early mammals are classified according to their teeth, since teeth are most commonly preserved as fossils. Types of teeth show us three early categories: gnawers, like modern mice and rats, that fed on plants; insect-eaters, with pointed side teeth for crushing; and those with varied teeth, who were able to eat a range of foods. The last group, the pantotheres, are the ancestors of modern mammals.

For the whole of the Mesozoic period, until 65 million years ago, mammals remained small, but they did develop into the separate groups that set the pattern for their future evolution. It was only at the end of the Mesozoic, with the extinction of the dinosaurs, that mammals began to flourish.

There are two principal types of mammal – marsupials and placentals. Marsupials give birth to under-developed offspring, which are kept in pouches in the mother's body where they feed and grow until they are ready to emerge. Placentals, which evolved later, give birth to fully developed offspring. In the present world, marsupials occur only in Australia, together with a few species in South America. Everywhere else placentals are the exclusive mammal type. But marsupial fossils have been found in Europe, North Ameria, South America and parts of Asia. Using this information we can trace the migration of mammals to every continent on Earth.

Marsupials first emerged on the northern continents, probably before placentals, and migrated both to the east and south, finding their way to Australia. Placental mammals, because their method of reproduction is more effective, generally replaced marsupials wherever they were in competition. But by the time advanced placentals reached South America, the land bridge to Australia via Antarctica had been cut off, leaving the marsupials of Australia in glorious isolation, removed from placental competition.

Ptilodus

Argyrolagus

Zalambdalestes

65 million years ago
Marsupials first emerged on the northern continents, probably a little before placental mammals. The earliest sites are in North America. They probably migrated eastwards into Europe and Asia and south through South America, across Antarctica and into Australia. There are no marsupials in New Zealand, which split from Australia 80 million years ago, so we can assume they arrived in Australia after that time.

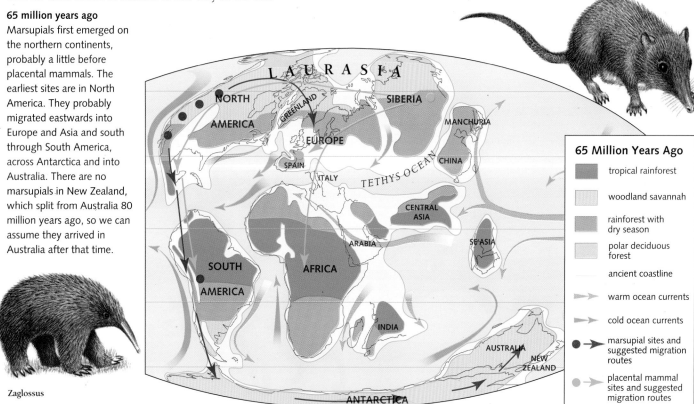

Zaglossus

65 Million Years Ago	
	tropical rainforest
	woodland savannah
	rainforest with dry season
	polar deciduous forest
	ancient coastline
	warm ocean currents
	cold ocean currents
	marsupial sites and suggested migration routes
	placental mammal sites and suggested migration routes

50 million years ago

The southern continents became isolated from the northern continents and from each other over the next 50 million years. During that time placental mammals underwent their own spectacular development. But by the time advanced placentals arrived in South America the land bridge to Australia had been cut off by continental movement, and also by the glaciation of Antarctica, which killed off the endemic life of that continent. It is not clear why early placental mammals did not cross over to Australia with the marsupials – there may have been some environmental conditions which discouraged them.

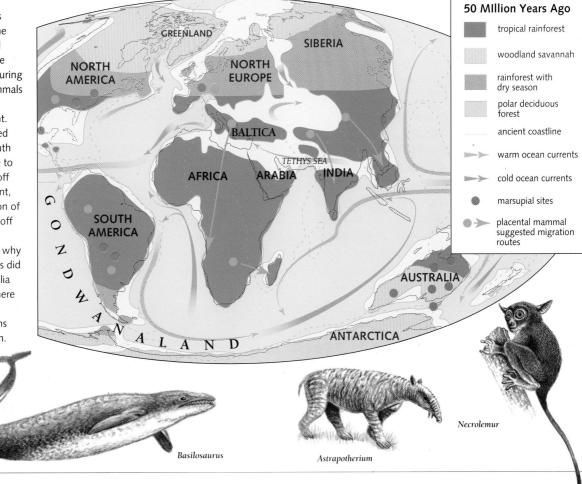

Basilosaurus

Astrapotherium

Necrolemur

25 million years ago

Marsupials existed in North America and Europe 25 million years ago, today they occur in Australia and South America only. With no competition from the placental mammals, marsupials evolved to fill a wide range of ecological niche. Australian fossil records include forms resembling placental dogs, cats, wolves, bears, squirrels, pigs, rodents.

Hesperocyon

Megaceros

Megistotherium

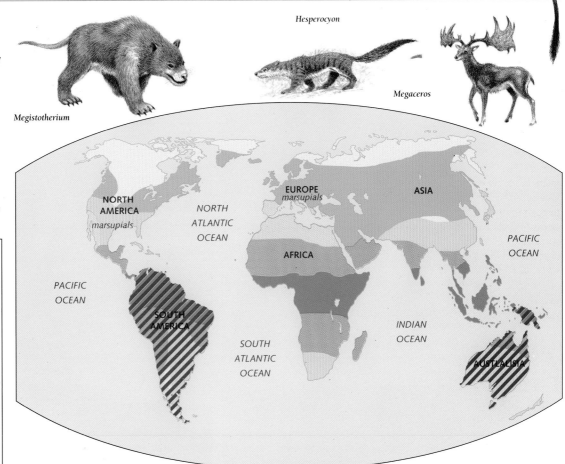

The Cordilleran Orogeny and the Rocky Mountains

Uplift of western North America

The focus of geological activity on the North American continent shifted from the east coast to the western margin during the Mesozoic era. The geological history of western North America is complex and fascinating. The westward movement of the craton has brought it into continual conflict with the oceanic plates that make up the floor of the Pacific Ocean. The Pacific plates themselves are moving in different directions, causing disturbances all around the region. These are

continuing to the present day, as we know from the eruption of Mt St Helens (*page 160*), and the earthquakes along the San Andreas fault (*page 162*).

The formation of western North America took place in a series of disturbances, the Cordilleran Orogeny, which began in the Late Jurassic period (150 million years ago) and lasted until about 40 million years ago. Since the end of the Cordilleran Orogeny further disturbances have occurred, but these have been on a smaller scale.

The Cordilleran Orogeny
The history of the Cordilleran Orogeny shows that the zone of disturbance moved from the west coast to the Rocky Mountains and back again, operating over a vast distance. This can only be explained by an extremely shallow angle of subduction – allowing the oceanic plate to cause friction and heating a long way under the margin of the continental plate.

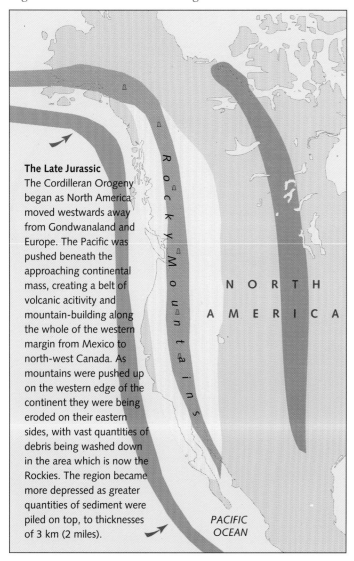

The Late Jurassic
The Cordilleran Orogeny began as North America moved westwards away from Gondwanaland and Europe. The Pacific was pushed beneath the approaching continental mass, creating a belt of volcanic acitivity and mountain-building along the whole of the western margin from Mexico to north-west Canada. As mountains were pushed up on the western edge of the continent they were being eroded on their eastern sides, with vast quantities of debris being washed down in the area which is now the Rockies. The region became more depressed as greater quantities of sediment were piled on top, to thicknesses of 3 km (2 miles).

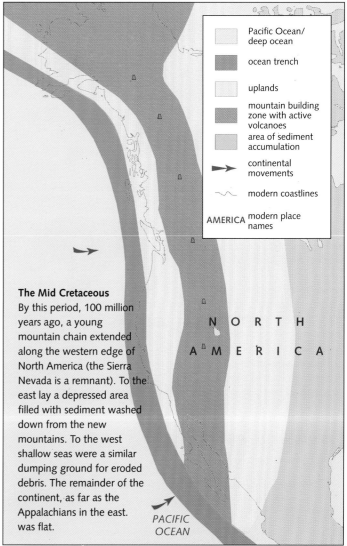

Pacific Ocean/ deep ocean

ocean trench

uplands

mountain building zone with active volcanoes

area of sediment accumulation

continental movements

modern coastlines

AMERICA modern place names

The Mid Cretaceous
By this period, 100 million years ago, a young mountain chain extended along the western edge of North America (the Sierra Nevada is a remnant). To the east lay a depressed area filled with sediment washed down from the new mountains. To the west shallow seas were a similar dumping ground for eroded debris. The remainder of the continent, as far as the Appalachians in the east, was flat.

Sierra Nevada, California
The coastal ranges and the Sierra Nevada (*right*) were formed in the second stage of the Cordilleran Orogeny. As the Pacific plate was subducted under western North America, off-shore island arcs came into collision with the mainland, causing mountain building.

Suspect terranes of western North America
Suspect terranes are pieces of crust that are formed elsewhere and which subsequently become attached to the edges of continental cratons. Continents grow either by volcanic activity at their margins, or by the uplift of sedimentary basins, or by the accretion of suspect terranes.

Geologists once thought that all the pieces of crust that became attached to continental margins were island arcs that formed off-shore. But it is now known that some suspect terranes travel thousands of kilometres on ocean plates before becoming attached to the edges of continents.

The western edge of North America is made up almost entirely of suspect terranes, mostly accreted during the final stages of the Cordilleran Orogeny.

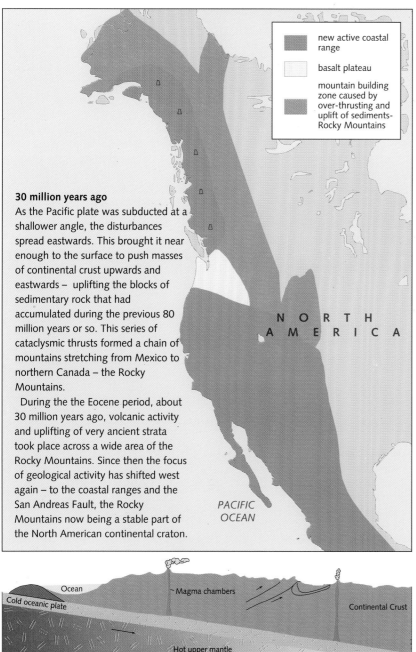

new active coastal range

basalt plateau

mountain building zone caused by over-thrusting and uplift of sediments- Rocky Mountains

30 million years ago
As the Pacific plate was subducted at a shallower angle, the disturbances spread eastwards. This brought it near enough to the surface to push masses of continental crust upwards and eastwards – uplifting the blocks of sedimentary rock that had accumulated during the previous 80 million years or so. This series of cataclysmic thrusts formed a chain of mountains stretching from Mexico to northern Canada – the Rocky Mountains.

During the the Eocene period, about 30 million years ago, volcanic activity and uplifting of very ancient strata took place across a wide area of the Rocky Mountains. Since then the focus of geological activity has shifted west again – to the coastal ranges and the San Andreas Fault, the Rocky Mountains now being a stable part of the North American continental craton.

THE WEST COAST: The suspect terranes

Rocky Mountain zone – sedimentation followed by uplift

rocks formed in present positions

suspect terranes containing Paleozoic rocks (greater than 250 million years old)

suspect terranes containing Mesozoic rocks (250 - 65 million years od)

The Himalayas
India in collision with Asia

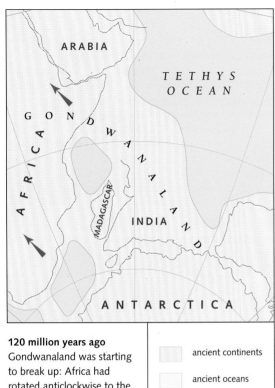

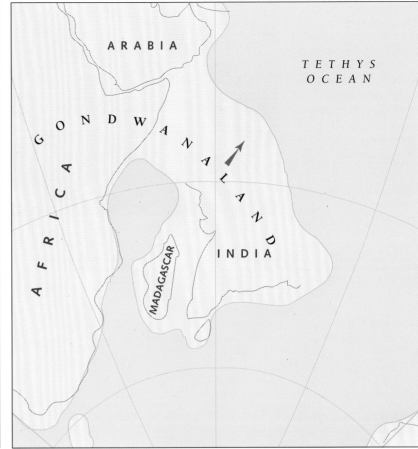

120 million years ago
Gondwanaland was starting to break up: Africa had rotated anticlockwise to the west, splitting from India and Antarctica, which were joined. Madagascar was still joined to India.

- ancient continents
- ancient oceans
- modern coastlines
- continental movements

Between 10 and 20 million years ago the Indian subcontinent collided with the continent of Asia. The Indian plate continued driving northward, penetrating about 2,000 kilometres (1,250 miles) into Asia. A piece of continental crust as wide as India and 2,000 kilometres deep was therefore displaced. It seems that the Himalayas are formed from slices of the old crust of northern India piled on top of one another in a dramatic example of overthrusting on a continental scale.

The effects of the collision also spread sideways, pushing China to the east, with slabs of continent moving along newly formed slip faults. The plateau of Tibet, a relatively flat area, was uplifted vertically to relieve the pressure of the horizontal movement. Gravity measurements reveal that the continental crust beneath the Himalayas is roughly twice as thick as the average. As well as crust being pushed upwards by horizontal pressure, it was also forced down, into the existing continent edges.

The Indian collision with Asia coincided with the African/European collision, creating the Alpine-Himalayan belt, a zone of mountain-building and earthquakes stretching from France to China.

85 million years ago
India had now split off from Africa and from Antarctica, and was moving northwards across the Tethys Ocean. Antarctica had moved off this map to the south, and Australia to the south-east.

The Himalayas (right)
The height and sharpness of the Himalayan mountain peaks is evidence of their comparative youth. They are the youngest and highest mountain chain in the world, with ten peaks of over 8,000 metres (26,000 ft). Sediment from the erosion of the Himalayas has filled the plains of northern India and is carried to the Bay of Bengal by the Ganges and Brahmaputra river systems.

50 million years ago to the present-day

By 50 million years ago India had moved north and rejoined a massive continental plate comprising Africa and Arabia. These continents remained locked together and continued to move northwards towards the great continental mass of Europe and central Asia. A smaller plate containing Southeast Asia was also moving north towards China.

The collision between these huge continental blocks happened over a long period, creating mountain chains and disturbances over thousands of kilometres. The climax of India's collision with Asia came in the last 10 million years, creating the Himalayas, the highest mountain range in the world.

Effects of the collision

The Himalayan Frontal thrust fault marks the edge of the old Indian plate driving into Asia. Some of the horizontal compression was translated into vertical motion by uplifting the Tibetan plateau. The remainder of the horizontal compression was relieved by pushing China to the east. A huge part of the Asia plate was forced sideways like toothpaste being squeezed out of a tube.

The Himalayan mountain chain seen here at Sagarmatha National Park, Nepal follows the northern boundary of the Indian plate. Mountain chains subside once the plate movements come to a stop. In some cases the crust behaves like a thick fluid if something is being pushed into it, wrinkles will appear, but once the movement stops the surface becomes flat again.

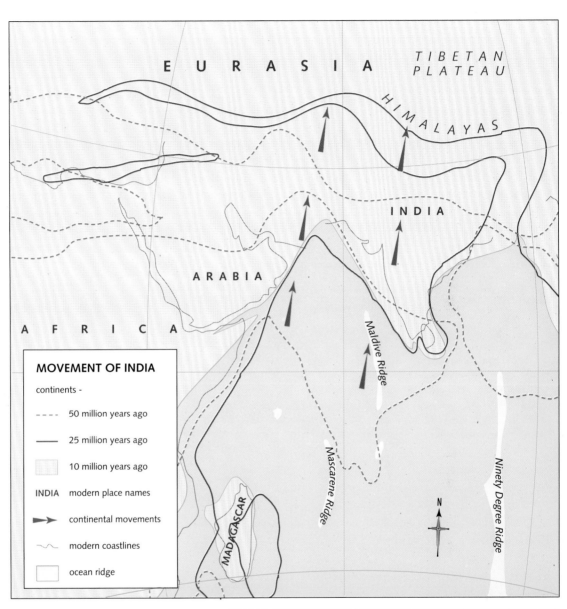

MOVEMENT OF INDIA

continents -

- - - - 50 million years ago

———— 25 million years ago

     10 million years ago

INDIA   modern place names

➤   continental movements

〜〜   modern coastlines

☐   ocean ridge

Australasia
The long separation from Gondwanaland

Australia, along with New Guinea and New Zealand, split from the remaining parts of the continent of Gondwanaland about 85 million years ago. These islands remained isolated from the rest of the world for 75 million years, developing a unique flora and fauna.

The mammals that had reached Australia before its separation from Antarctica were all marsupials; animals that give birth to dependent young, which are kept in pouches in the mother's body for the earliest part of their lives. In other parts of the world these were gradually replaced by placental mammals, which give birth to fully developed young. Australian marsupials therefore developed to occupy the ecological niches occupied by placentals elsewhere. The small group of egg-laying mammals (monotremes) is unique to Australasia, because it has been entirely replaced elsewhere. Large flightless birds have also thrived in the absence of placental mammals – particularly in New Zealand, where there are no indigenous mammals at all apart from bats.

In the last 5 to 10 million years, as Australasia has continued to move northwards, its northern margin has come in contact with the islands of Southeast Asia. This has ended the isolation of Australasia, but the faunas and floras of the two continents are still distinct. Some species have crossed over to Australia from Southeast Asia and vice versa, but it will take millions more years for the distinction between the two continental faunas and floras to fade.

The separation of Australia
By the Cretaceous period, 120 million years ago, Gondwanaland was breaking up. Australia, Tasmania and New Guinea were still attached to Antarctica – the centre of Gondwanaland. New Zealand was also joined to Antarctica, but its position relative to Australia was very different from the present. The north-east margin of India was still joined to the west of Australia, while Southeast Asia had split off from the north of Australia 30 million years earlier and was moving northwards.

By 85 million years ago the Australian continent had split from Antarctica along the whole of its southern margin, and moved north. India and Southeast Asia had moved away, leaving Australia relatively isolated. New Zealand was still attached to Australia.

By 50 million years ago the gap between Australia and Antarctica had widened, as they both moved in opposite directions. By now the migration of land animals between the southern continents had become virtually impossible. New Zealand was now separate from Australia and animal migration between the two ceased.

New Zealand has now rotated clockwise to create the full opening of the Tasman Sea.

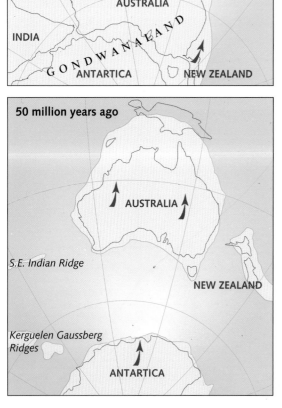

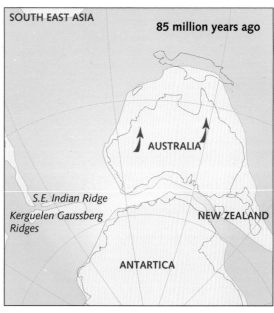

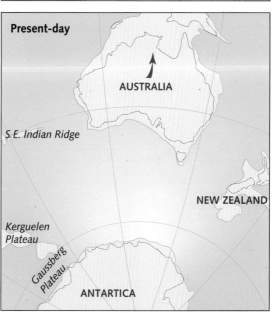

THE SEPARATION
OF AUSTRALASIA

	ancient continents and continental shelf
BALTICA	ancient place names
~~	modern coastlines
➤	continental movements
NEW ZEALAND	modern place names

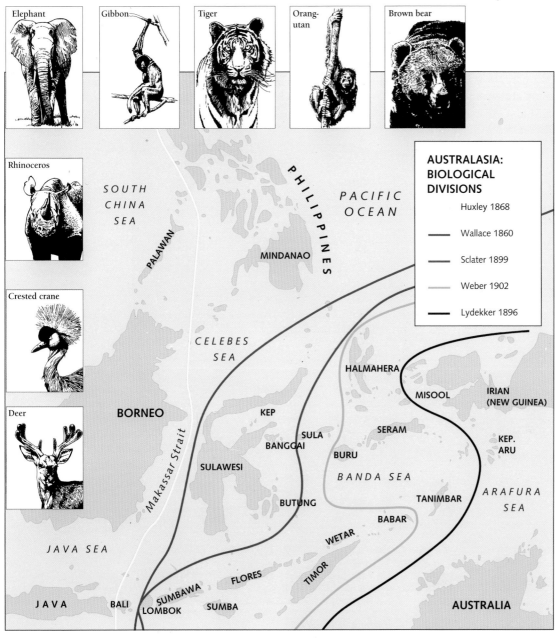

Elephant

Gibbon

Tiger

Orang-utan

Brown bear

Koala bear

Rhinoceros

SOUTH
CHINA
SEA

PHILIPPINES

PACIFIC
OCEAN

PALAWAN

MINDANAO

**AUSTRALASIA:
BIOLOGICAL
DIVISIONS**

Huxley 1868

Wallace 1860

Sclater 1899

Weber 1902

Lydekker 1896

Crested crane

CELEBES
SEA

HALMAHERA

MISOOL

IRIAN
(NEW GUINEA)

KEP

SULA

SERAM

KEP.
ARU

BANGGAI

BURU

Deer

BORNEO

SULAWESI

BANDA SEA

ARAFURA
SEA

Makassar Strait

BUTUNG

TANIMBAR

BABAR

JAVA SEA

WETAR

JAVA   BALI   SUMBAWA   SUMBA

LOMBOK

FLORES   TIMOR

AUSTRALIA

Duck-billed
platypus

Kangaroo

Asian and Australian fauna
The drawings above show
examples of Asian (*top and
left of map*) and Australian
(*right of map*) mammals and
birds, which have not yet
crossed the Wallace Line.
Some plant and animal
groups have naturally
migrated, and some (for
example rabbits and dogs)
have been brought to
Australia by humans.

The Wallace Line
The movement of Australasia brought about one of the
most famous discoveries in scientific history. In the 1850s
the biologist Alfred Wallace noticed that bird populations
on neighbouring islands in the East Indies belonged to two
separate groups with almost no movement between them.
One group had affinities with the birds of Asia, another
with Australian types. The same was found to be true for
mammals and molluscs, though the boundary between
these groups was in a slightly different place.

Wallace showed that the presence of different groups
could be explained by separate evolution. But he did not
realize that the reason for this was the long separation of
Australia and Asia, which have only relatively recently drifted together. Subsequent biologists have drawn line
between the different groups of Asian and Australian fauna
and flora in slightly different places (*see map above*).

Mount Ngauruhue
This classic volcanic cone
(*left*) on North Island is
evidence of tectonic activity
caused by a plate boundary
which runs straight through
New Zealand.

Corals

Changing sea levels and reef-building

The importance of corals in understanding the Earth's history is out of all proportion to the size of the individual animals. But because of their tendency to join together to form massive reef structures, and their sensitivity to environmental conditions, corals are a key group of animals.

Corals belong to the phylum Cnidaria, which also contains the sea anemones and the jellyfish. The simple sack-like body of these animals has a wall of two-layered tissue and one cavity (a mouth) usually surrounded by a ring of tentacles. While many cnidarians are soft-bodied and are rarely preserved as fossils, corals are a spectacular exception.

Corals are like sea-anemones in form, though they produce an underlying skeleton that raises them above the sea bed. They form either a simple conical skeleton, secreted and used by a single individual, or they may form complex compound skeletons. Compound corals may be of the branching type, or massive. Massive types form colonies where the individual corallites join together and lose their intervening walls. These can form the huge structures that we know as coral reefs. These reefs are made up of the skeletons of millions of individual corals – some living and some dead.

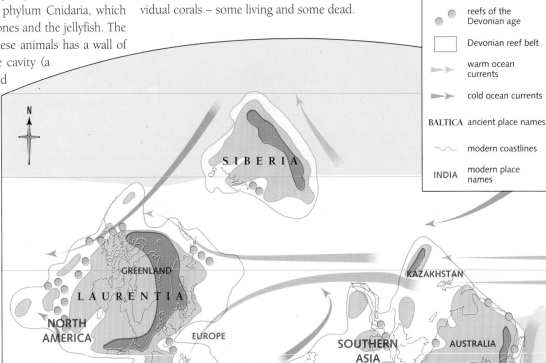

DEVONIAN CORALS

- ancient continents
- ancient continental shelf
- ancient mountain chains
- reefs of the Devonian age
- Devonian reef belt
- warm ocean currents
- cold ocean currents
- BALTICA ancient place names
- modern coastlines
- INDIA modern place names

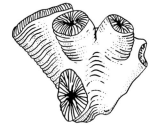

Devonian Corals

The Devonian period was the first time in the Earth's history that organic reefs of significant size were built – by corals and other animals.

Coral forms

Corals appear in a variety of forms both individual (*left*), tubular (*right*), and in colonies (*above left*) and (*above right*).

Devonian reefs

The discovery of ancient coral reefs in the rocks of continents well away from the tropics, gave an early impetus to the theory of continental drift. Reef-building first became extensive in the Devonian period – 400 million years ago. The map shows that Devonian reefs were restricted to the same tropical areas as modern reefs.

Throughout the geological record reefs are major sites of deposition of carbonate rocks. The individual coral polyps secrete walls made of carbonate minerals. As conditions change the reefs may become 'dead'. A huge carbonate structure is then left behind, which may become compressed into rock as more sediment is piled on top. In this way rock formations are made with reefs integrated into them. Reefs can reach remarkable depths and, when they are uplifted by subsequent events, can become massive rock formations. There are many exmples of reefs forming parts of mountain chains. One of the most spectacular is in the Guadelupe Mountains of Texas.

The world's coral reefs

There are severe restrictions on where corals and other reef-builders grow successfully. These limitations make reefs unusually useful in reconstructing ancient environments.

Scleratinian corals need shallow water within the depth of sunlight penetration – roughly 20 metres (66 ft). They need sunlight as they live in symbiosis with small green algae living in the translucent coral tisssue and need light to live.

The seawater needs to be between 20 and 30°C and of normal salinity. A firm anchorage on a hard sea bottom is required and there must be no muddy sediment. Shallow, well-lighted warm water is therefore required and this restricts abundant coral reef growth to the tropics.

Corals and other reef-building animals grow best where currents wash over them. By studying ancient reefs it is therefore possible to determine tidal and current movements as well as climate and depth.

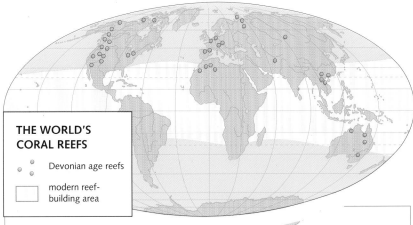

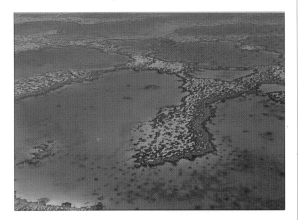

The Barrier Reef

This 1,600 km (1,000 mile) complex of reefs forms a gigantic natural breakwater off the north-east coast of Australia and is easily the largest coral structure in the world (*above*).

Red Sea coral

This modern coral polyp (*right*) is a solitary form, not a reef-builder, though polyps often live in communities.

Coral polyp

Each individual polyp (*right*) has tentacles for drawing food into its mouth. It has a stony external skeleton.

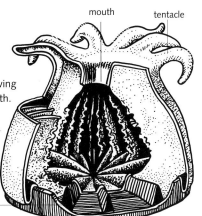

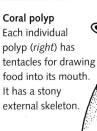

mouth   tentacle

stony skeleton

THE WORLD'S CORAL REEFS

Devonian age reefs

modern reef-building area

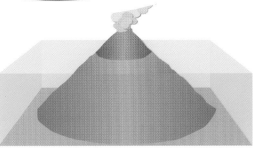

An island develops as volcanoes erupt along underwater plate boundaries.

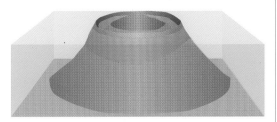

Corals begin to form a circular reef around the margins of the volcanic island.

As the volcano sinks – either through rising sea levels or a subsiding crust – the reef grows higher and forms a lagoon.

The inactive volcano sinks leaving the circle of the slightly submerged reef. The flat floor of the coral island lagoons is due to erosion and deposition.

Evolution of a coral atoll

Charles Darwin was the first scientist to make a study of the development of coral islands and atolls. On his historic voyage aboard the *Beagle* from 1831 to 1836 he recognized the different topographies of many tropical islands as stages in a similar evolving pattern. This is a fascinating example of how Darwin was seeking to explain natural phenomena as evolutionary events so early in his scientific career. Darwin's theories of coral island development are still widely accepted – a remarkable achievement in such a rapidly changing science.

The Hawaiian Islands

A 'hot spot' on the Earth's crust

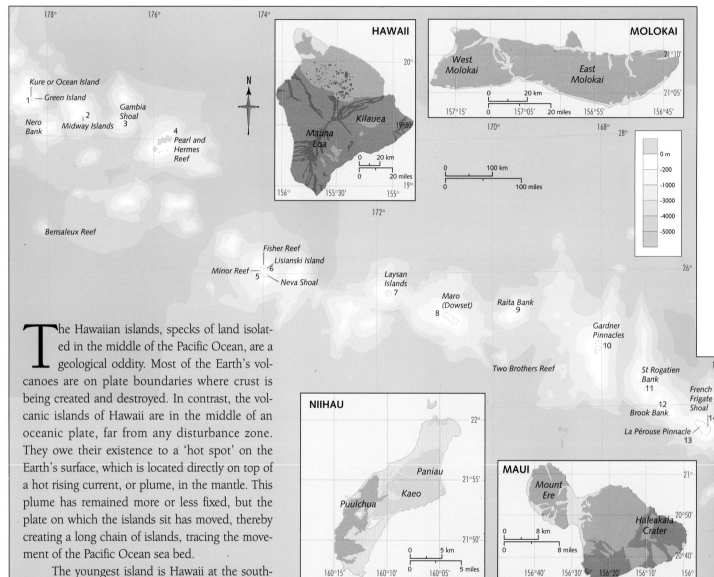

HAWAII

Mauna Loa

Kilauea

MOLOKAI

West Molokai

East Molokai

0 m
-200
-1000
-3000
-4000
-5000

Kure or Ocean Island
Green Island
Nero Bank
Midway Islands
Gambia Shoal
Pearl and Hermes Reef

Bensaleux Reef

Fisher Reef
Lisianski Island
Minor Reef
Neva Shoal

Laysan Islands

Maro (Dowset)

Raita Bank

Gardner Pinnacles

Two Brothers Reef

St Rogatien Bank

French Frigate Shoal

Brook Bank

La Pérouse Pinnacle

NIIHAU

Paniau
Kaeo
Puulchua

MAUI

Mount Ere
Haleakala Crater

OAHU

Schofield Plateau
Honolulu

KALOOLAWE

Lua Makika

The Hawaiian islands, specks of land isolated in the middle of the Pacific Ocean, are a geological oddity. Most of the Earth's volcanoes are on plate boundaries where crust is being created and destroyed. In contrast, the volcanic islands of Hawaii are in the middle of an oceanic plate, far from any disturbance zone. They owe their existence to a 'hot spot' on the Earth's surface, which is located directly on top of a hot rising current, or plume, in the mantle. This plume has remained more or less fixed, but the plate on which the islands sit has moved, thereby creating a long chain of islands, tracing the movement of the Pacific Ocean sea bed.

The youngest island is Hawaii at the southeastern end of the chain. The islands grow progressively older to the north-west – a trend which is continued in the Emperor seamounts, a chain of former volcanic islands, which extends on from the Hawaiian group.

Only Hawaii and Maui have volcanoes that are still active. Light basalt lava in the upper mantle finds its way up through a fissure in the crust and collects in chambers beneath the volcanoes. In Hawaii, lava is generally released gradually through fissures. but sometimes it is held under the volcano, gradually building up pressure until it explodes through the overlying rock.

The Hawaiian island chain

The newest island of the chain, Hawaii, is 6 million years old. It is still situated over the hot plume in the mantle and its two great volcanoes, Mauna Loa and Mauna Kea, are still active. The ocean is 6,000 metres (20,000 ft) deep, and each island was built up from eruptions on the sea floor in just a few million years – an unusually rapid rate of eruption. The youngest islands consist almost entirely of volcanic rock, since there has not been enough time for erosion and deposition of sediment to take place. The older islands show progressively more sediment, though this is very dependent on the topography of the individual islands. More recent lava flows also follow natural contours, so that they too are deposited in valleys and basins.

CROSS-SECTION THROUGH HAWAIIAN ISLAND CHAIN *(numbers refer to map)*

The hot plume

The postulated hot plume under the Hawaiian islands pushes magma up through the surface of the Pacific plate. The plume remains stationary, while the Pacific plate moves slowly over it. The Hawaiian islands become younger towards the south-east, and a line of older sea mounts, (islands which no longer reach the ocean surface) stretches away in a northwesterly direction.

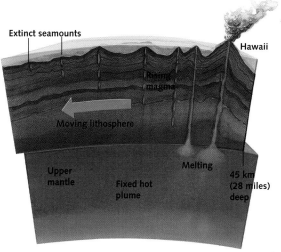

Extinct seamounts

Hawaii

Rising magma

Moving lithosphere

Upper mantle

Fixed hot plume

Melting

45 km (28 miles) deep

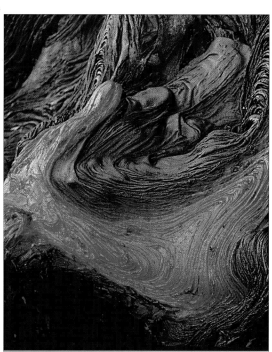

Kilaeua

The active volcano of Kilauea (*right*) on Hawaii. It continually extrudes molten lava (*far right*). The last great eruptions were in 1959 and 1960.

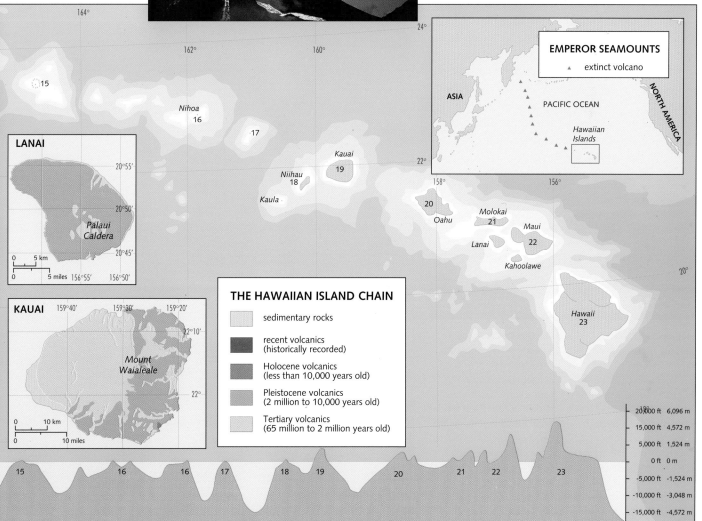

LANAI

20°55'

20°50'

Palaui Caldera

20°45'

0   5 km

0   5 miles   156°55'   156°50'

KAUAI

159°40'   159°30'   159°20'

Mount Waialeale

22°10'

22°

0   10 km

0   10 miles

164°

162°

160°

Nihoa 16

·17

Kauai 19

Niihau 18

Kaula

15

20°

EMPEROR SEAMOUNTS

▲ extinct volcano

ASIA

PACIFIC OCEAN

NORTH AMERICA

Hawaiian Islands

24°

22°

158°   156°

20

Oahu

Molokai 21

Maui 22

Lanai

Kahoolawe

20°

Hawaii 23

THE HAWAIIAN ISLAND CHAIN

sedimentary rocks

recent volcanics (historically recorded)

Holocene volcanics (less than 10,000 years old)

Pleistocene volcanics (2 million to 10,000 years old)

Tertiary volcanics (65 million to 2 million years old)

20,000 ft   6,096 m

15,000 ft   4,572 m

5,000 ft   1,524 m

0 ft   0 m

-5,000 ft   -1,524 m

-10,000 ft   -3,048 m

-15,000 ft   -4,572 m

15   16   16   17   18   19   20   21   22   23

The Northern North Atlantic
From Iceland to the Arctic Basin

Northern Europe and North America had been joined together since the collisions of the Caledonian Orogeny 400 million years ago. About 50 million years ago, during the Early Tertiary period, the two continents began to split apart once more.

During the break-up of Pangea, the central Atlantic had been formed by the separation of the ancient continents of Gondwanaland and Laurentia. The sea floor spreading had then extended south to separate South America from Africa. The opening of the northern Atlantic joined the central Atlantic basin to the Arctic Ocean basin, forming a zone of sea floor spreading reaching from the Antarctic to the Arctic. During this process a portion of Siberia split off from the main continent to form the Lomonosov Ridge – an underwater ridge which runs across the North Pole splitting the Arctic basin in two.

The mid-Atlantic ridge approaches the ocean surface in the North Atlantic, at the Reykjanes Ridge just south of Iceland. The island of Iceland is itself part of the mid-Atlantic ridge. Ocean crust, in the form of basalt lava, is extruded from fissures running roughly north-south through the island. All of the rock on Iceland are less than 20 million years old – there are virtually no sediments. In the interior of the island geysers are formed by ground water evaporating on contact with the hot magma lying just beneath the surface.

The North Atlantic 50 million years ago
The Arctic basin was completely enclosed by the surrounding continents. On the northern edge of Europe a large area of continental shelf, the Barents Sea, bordered the deep Arctic basin. Europe, Greenland and North America were part of the same landmass, with the first signs of splitting between north-west Europe and Greenland.

ALASKA

CANADIAN ARCTIC BASIN

SIBERIA

BARENTS SEA

GREENLAND

EUROPE

ancient continents and continental shelf

deep ocean

modern coastlines

SIBERIA modern place names

Hot rocks
Because of Iceland's position on top of a rising convection current in the Earth's mantle, hot rocks are very near the surface. When ground water percolates down through cracks in the crust it eventually meets these rocks. It is instantly turned into steam and ejected back up to the surface. This can happen in bursts, as in a geyser, or continually (*above*).

The Giant's Causeway
As north-west Europe and Greenland were torn apart fissures opened up in the Earth's crust. Basalt was pushed up into these fissures from the hot mantle. As the basalt cooled and contracted it sometimes developed cracks. At Giant's Causeway in Ireland (*right*) the cracks split the basalt into spectacular hexagonal columns.

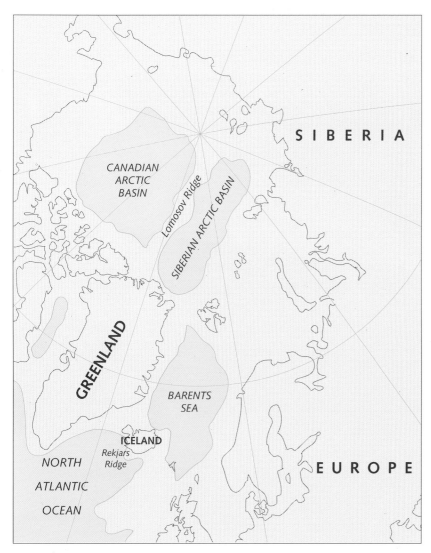

The North Atlantic 25 million years ago

Deep ocean had opened up between Greenland and Norway and the British Isles. The mid-Atlantic ridge now extended from the tropics to the edge of the Arctic basin, and was the source of the sea floor spreading that pushed the continents apart.

In the Arctic a new basin developed to the north of Siberia. This pushed a strip of the continental shelf away to the north as it developed. This strip became the Lomonosov ridge.

The present day

The northern North Atlantic is now linked to the new basin on the Siberian side of the Arctic. This ocean continues to widen as Europe moves away from Greenland, and Siberia separates from the Lomonosov ridge. Iceland sits on the mid-Atlantic ridge in a section where it is near to the ocean surface.

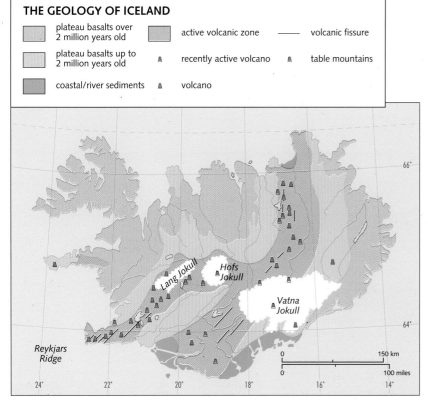

THE GEOLOGY OF ICELAND

- plateau basalts over 2 million years old
- plateau basalts up to 2 million years old
- coastal/river sediments
- active volcanic zone
- recently active volcano
- volcano
- volcanic fissure
- table mountains

The peculiar case of Iceland

Iceland is a geological freak – a piece of land made of oceanic crust. It sits on top of the mid-Atlantic ridge, and is entirely made up of rock extruded from the fissures in the ridge. Iceland did not exist 100 million years ago. Analysis of the precise ages of the rocks on Iceland has shown that the centre of sea floor spreading on the ridge shifted by about 150 kilometres (93 miles) around 1.5 million years ago. The southern centre of spreading seems to have moved back to the east again. Iceland remains volcanically active. Activity includes hot geysers in the central parts of the island.

In the 1973 eruption of Heimaey, a volcanic island off the coast of Iceland huge quantities of cinders and ash were blown onto houses, causing roofs to collapse. The lava flows created a bigger island with a better harbour.

The Alps and the Mediterranean Basin
Africa and Arabia collide with Europe and Asia

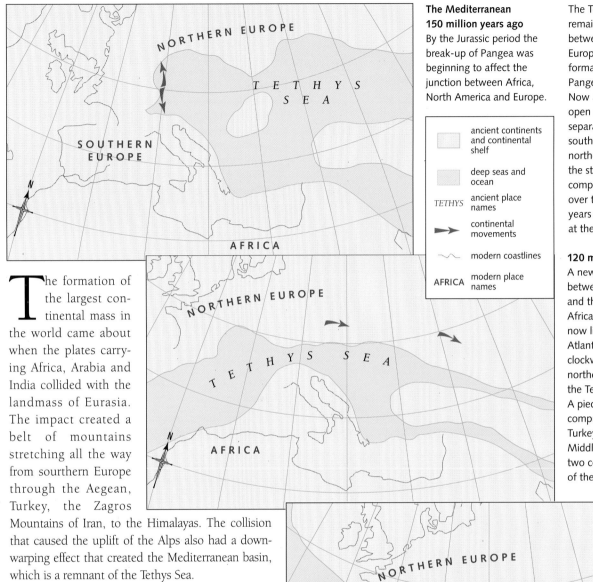

**The Mediterranean
150 million years ago**
By the Jurassic period the break-up of Pangea was beginning to affect the junction between Africa, North America and Europe.

	ancient continents and continental shelf
	deep seas and ocean
TETHYS	ancient place names
➤	continental movements
∿	modern coastlines
AFRICA	modern place names

The Tethys Sea had remained as an inlet between North Africa and Europe throughout the formation of the great Pangean supercontinent. Now a new split began to open up at its western end, separating Africa and southern Europe from northern Europe. This was the start of a series of complex movements lasting over the next 150 million years – and still continuing at the present time.

120 million years ago
A new ocean appeared between northern Euopre and the northern margin of Africa. The Tethys Sea was now linked to the new Atlantic Ocean, though the clockwise motion of northern Europe narrowed the Tethys at its eastern end. A piece of continent comprising present-day Turkey and parts of the Middle East lay between the two continents in the middle of the Tethys.

The formation of the largest continental mass in the world came about when the plates carrying Africa, Arabia and India collided with the landmass of Eurasia. The impact created a belt of mountains stretching all the way from southern Europe through the Aegean, Turkey, the Zagros Mountains of Iran, to the Himalayas. The collision that caused the uplift of the Alps also had a downwarping effect that created the Mediterranean basin, which is a remnant of the Tethys Sea.

The process known as the Alpine Orogeny took place in the Miocene period, about 10 million years ago. But like many of the geological events that shaped our present Earth, its origins can be traced back to the break-up of the supercontinent of Pangea, 150 million years ago. Since then Europe and Africa have moved apart and collided again in several episodes. Central Europe and the western Mediterranean is now a geologically stable region, as the two continents are joined at this point with only a weak plate boundary between them. But further east movement between Africa, Europe, Arabia and Anatolia continues sporadically. This active zone, which stretches from southern Italy through Greece, Turkey, and the Caspian region to Iran is the site of earthquakes. The active volcanoes of Vesuvius and Etna are evidence of continuing movements beneath the crust in this unstable region.

80–85 million years ago
Towards the end of the Cretaceous a spreading zone in the Bay of Biscay rotated Spain into its present position. The first signs of uplift of the Pyrenees occurred at this time. The Tethys Sea was once again cut off from the Atlantic. A zone of spreading had developed between Spain and North Africa.

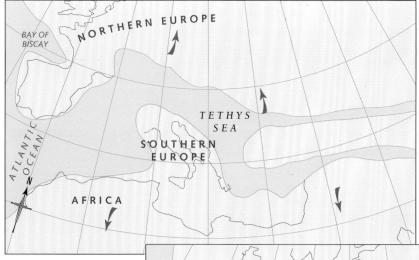

50 million years ago
In the early Tertiary period the last great separation of Africa and Europe occurred. Africa moved south with Italy and southern Europe attached. There was clear ocean between northern Europe, now including Spain, and the continent of Africa. High ground in present-day Turkey kept this area above sea level throughout this process.

25 million years ago
The African plate had now moved north in full collision with central and northern Europe. This huge continental collision lasted for millions of years and profoundly affected the topography and geology of the region. Italy was at the northern edge of the African plate and was driven into central Europe, causing the uplift of the Alpine region. The effect on the area between north Africa and southern Europe was downwarping (buckling downwards). This produced a great basin in the middle of the two continents, which was eventually to fill with seawater to form the Mediterranean. By this time the geography of the continent was recognisably similar to the present, though a remnant of the Tethys Sea lingered along the northern edge of Turkey.

Mount Etna
Italy has been part of the continent of Africa since Mesozoic times. Now that its northward motion is blocked, relative motion between Italy and Africa is causing volcanic activity, as at Mount Etna, Sicily's active volcano.

Alpine peaks
The Alps are a young chain of mountains, with sharp, high peaks. They are being eroded by glaciation and water.

10 million years ago
The Balkan region and the eastern Mediterranen were still mobile, while the Alps were still being pushed upwards. At present the Red Sea is the centre of sea floor spreading, pushing the Arabian peninsula north-westwards away from Africa. The continued motion of India into Asia keeps active those thrust faults that run through present-day Iran. This, combined with the continuing motion of Africa northwards makes the Eastern Mediterranean an extraordinarily complex tectonic region.

Africa continues to move north at the present time, but Europe stays where it is, creating a tension which is partly relieved by earthquakes, and partly by sections of continent being pushed sideways.

The Andes
Mountain chains and metal ores

The Andes region is the classic example of a subduction zone. The Nazca ocean plate has been pushing eastwards under the western edge of South America for several million years at the very fast rate of 15 to 18 cm (6-7 in) per year. This has created the Peru-Chile ocean trench, which reaches depths of 6,000 metres (20,000 ft), and the Andes whose highest peak, Aconcagua, stands at nearly 7,000 metres (22,965 ft).

The heat caused by the subduction of the Nazca plate, as well as creating volcanic mountains, has made the Andes among the world's richest regions for metallic ores. As the Nazca plate descends beneath the continent, it heats up – both through friction with the overlying continental mass,

and contact with the hot mantle beneath. As the temperature rises, different minerals in the rocks of the plates melt. Various minerals melt at different temperatures and are dissolved in the hot fluids that rise up through the crust, which cool as they near the surface, depositing the minerals in veins in the rock. Tin, copper, iron, lead, zinc, silver, gold and tungsten are all found in the Andes in economic quantities, and all as a result of heat from the subduction zone beneath.

Further east a different type of mineral deposit is formed as eroded material is washed down off the newly-formed mountains. Minerals are concentrated into significant deposits by the action of the groundwater on the sediments.

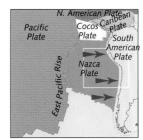

Shallow subduction
The Nazca plate which is being pushed rapidly eastwards by the spreading at the East Pacific Ridge, is still warm and buoyant when it reaches the western edge of the South American plate. Being light, its angle of descent into the mantle is relatively shallow. It is still near the surface, and still generating heat by friction with the underside of the South American continent, affecting the overlying continent for about 800 km (497 miles) inland. This explains the remarkable width of the Andes mountain chain.

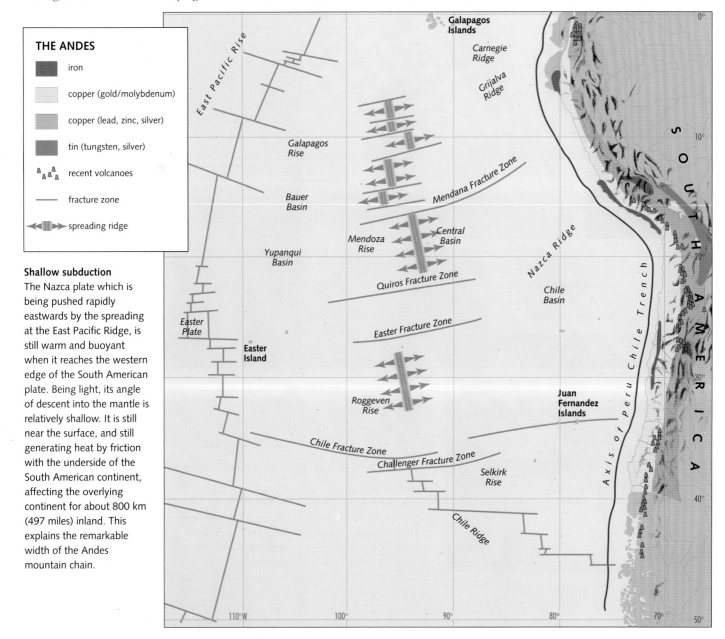

THE ANDES

- iron
- copper (gold/molybdenum)
- copper (lead, zinc, silver)
- tin (tungsten, silver)
- recent volcanoes
- fracture zone
- spreading ridge

Metal zones in the central Andes

The deeper the Nazca ocean plate descends beneath the South American continent the hotter it gets. As it descends the different minerals in the rocks of the plates melt. First to melt are those minerals containing iron (Fe) and manganese (Mn). Then a group containing copper (Cu), molybdenum (Mo) and gold (Au) melts out of the mixture. Then comes a group containing more copper minerals, minerals of lead (Pb) and zinc (Zn) and also silver (Ag). The final group to melt are minerals of tin (Sn) and tungsten (W) and more silver. This differential melting is reflected in the pattern of ore belts in the mineral-rich central zone of the Andes.

Open-cast mining, Peru
Open-cast mines like this one at La Oroya (left), are replacing traditional deep mines, as the richest ones are worked out, leaving low-grade ores to be dug out from near the surface.

Patagonian Andes
The Andes stretch for over 10,000 km (6,000 miles) from Colombia in the northern hemisphere to Patagonia (above), only 1,500 km (1,000 miles) from the Antarctic Circle.

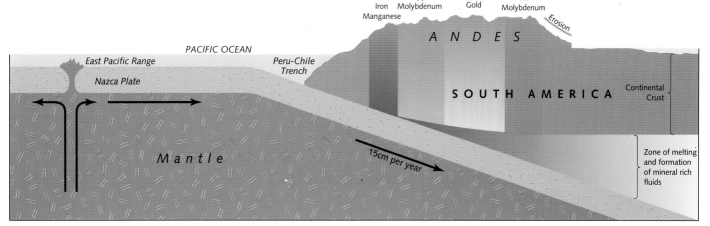

Azurite
Azurite is a deep-blue copper ore with a complex carbonate and hydroxide composition. It usually occurs in the upper parts of copper deposits, where more oxidation has taken place. Chemical formula is $Cu_3(CO_3)_2(OH)_2$.

Chalcopyrite
A copper iron sulphide with the chemical formula $CuFeS_2$, Chalcopyrite is the most important copper ore.

Gold
Gold does not react with other compounds, and this gives it its shine. It also means that gold almost always occurs as native metal – either in nuggets, grains or leaves. Though rare, gold is widely distributed in the rocks of the crust.

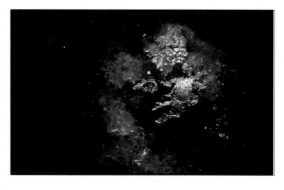

The Japanese Islands
Disturbance zones in the west Pacific

The Japanese islands lie in an arc off the eastern edge of the Asian continental mass. Japan is thought to have been torn away from the Asian continent in the mid-Cenozoic period, when the Pacific plate made a major change in direction. It switched from moving from south to north to moving south-east to north-west. This was accompanied by major worldwide tectonic changes, and there is some evidence that other western Pacific island arcs, such as New Zealand, were torn away from their parent continents at that time.

Japan is subject to continuous earthquakes and volcanic acitivity because of its position. As well as sitting on a subduction zone, it is also at the junction of two pieces of the massive Pacific plate, which are moving in slightly different directions. Understanding the geology of Japan is an important element in being able to predict the occurrence of further earthquakes and eruptions.

Mount Fuji
The most famous mountain in Japan is an active volcano lying on the Fossa Magna. This junction between two pieces of the Pacific plate, which runs across Honshu, is Japan's most active vocanic and earthquake zone.

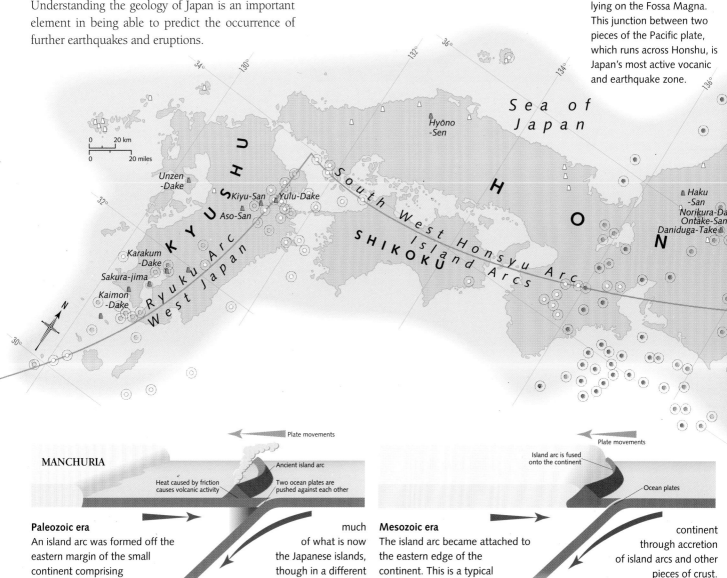

Paleozoic era
An island arc was formed off the eastern margin of the small continent comprising present-day Manchuria. This arc contained much of what is now the Japanese islands, though in a different configuration.

Mesozoic era
The island arc became attached to the eastern edge of the continent. This is a typical process leading to the gradual growth of the continent through accretion of island arcs and other pieces of crust.

The island arcs of Japan

Many of the world's active subduction zones and island arcs are located in the west and particularly the south-west Pacific region. The Marianas trench was formed where the Pacific plate was subducted beneath the Philippine plate at the same time as Japan.

The view of Japan as an island arc is a simplified one. In fact five separate arcs have been identified, each with their own tectonic history. They are grouped into two systems – West Japan and East Japan. These correspond to two major subduction zones as the Pacific plate is pushed beneath Japan and Asia. It is possible that the original Japanese island arc has been 'bent' into its present shape by various complex plate movements.

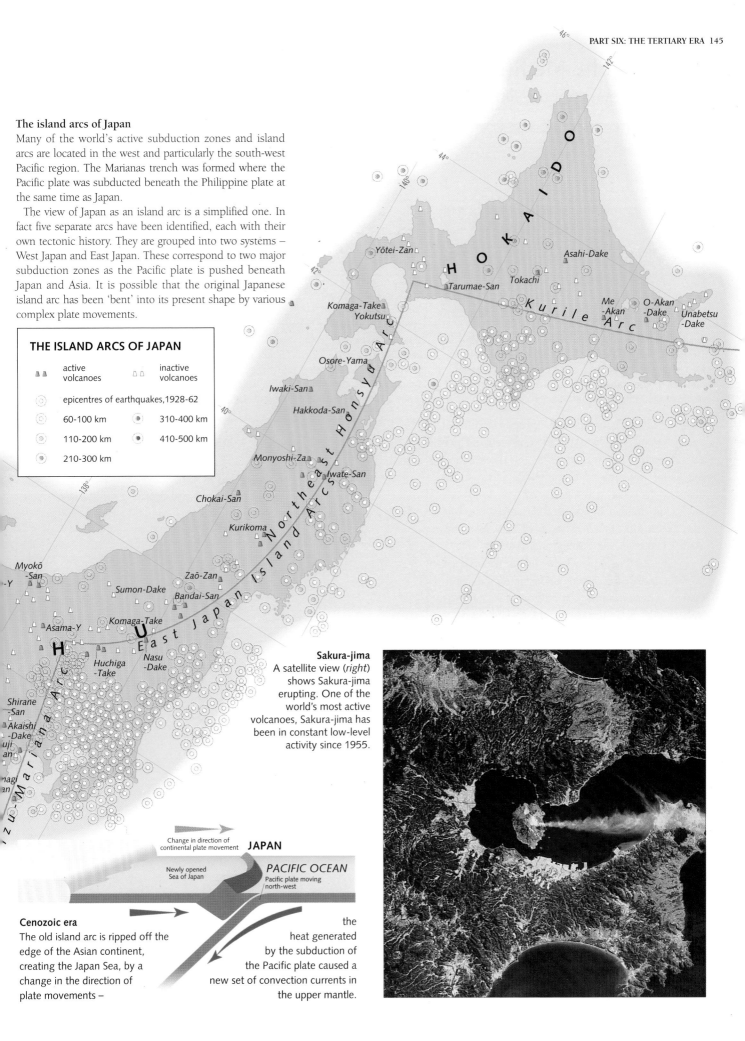

THE ISLAND ARCS OF JAPAN

▲ ▲ active volcanoes		△ △ inactive volcanoes
◉ epicentres of earthquakes, 1928-62		
◉ 60-100 km		◉ 310-400 km
◉ 110-200 km		◉ 410-500 km
◉ 210-300 km		

Sakura-jima
A satellite view (*right*) shows Sakura-jima erupting. One of the world's most active volcanoes, Sakura-jima has been in constant low-level activity since 1955.

Cenozoic era
The old island arc is ripped off the edge of the Asian continent, creating the Japan Sea, by a change in the direction of plate movements – the heat generated by the subduction of the Pacific plate caused a new set of convection currents in the upper mantle.

Change in direction of continental plate movement **JAPAN**

Newly opened Sea of Japan

PACIFIC OCEAN
Pacific plate moving north-west

Flooding the Mediterranean
The last remnants of the Tethys Sea

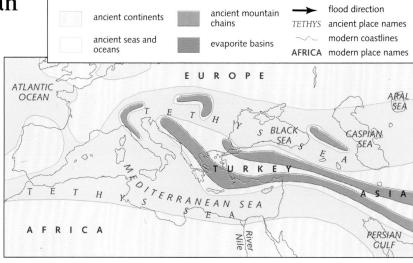

ancient continents

ancient seas and oceans

ancient mountain chains

evaporite basins

→ flood direction

TETHYS ancient place names

modern coastlines

AFRICA modern place names

Straits of Gibraltar
The first flooding of the Mediterranean probably came via southern France, and from the Tethys Sea to the east. Recent flooding of the Mediteranean has been via the Straits of Gibraltar, which connect it to the Atlantic.

A s Africa collided with Europe the ancient Tethys Sea, which had separated the two continents, was squeezed out of existence. But basins were created in the middle of the continents by the same movements, and were filled with water and drained in successive changes over the past 20 million years. They presently form the Mediterranean, Black, Caspian and Aral Seas.

The eastern edge of the Tethys was cut off from the Indian Ocean early on in the Alpine disturbances. Then, around 7–6 million years ago, the western edge was cut off from the Atlantic, and the Mediterranean dried out like a huge salt flat. Calculations suggest that the drying out of the whole Mediterranean would create a salt layer 100 metres (330 ft) thick on the sea bottom, but drilling has shown layers up to 2,000 (6,660 ft) metres thick. It seems that there were repeated intermittent reconnections to the Atlantic – possibly 40 refillings and evaporations took place in only 2 million years. It seems likely that this flooding, once thought to be through the Straits of Gribraltar, took place through the lowlands of south-west France.

At this time animals migrating from Africa to Europe are known to have walked across the basin. About 5 million years ago the Straits of Gibraltar opened, allowing the waters of the Atlantic to flood in and fill the Mediterranean once again. At the height of the Pleistocene ice ages the drop in world-wide sea levels separated the Black Sea from the Mediterranean on numerous occasions. They are now re-united, but the Caspian and Aral Seas, also remnants of the Tethys Sea, have been isolated for the past 3 million years.

20 million years ago
New island arcs and mountain chains were being formed between northern Europe and the approaching continent of Africa. The Tethys Sea was split into two branches by a mountain chain running through present-day Turkey and the Balkans. The link with the Atlantic and relatively high sea levels kept the Mediterranean region of the Tethys filled with seawater.

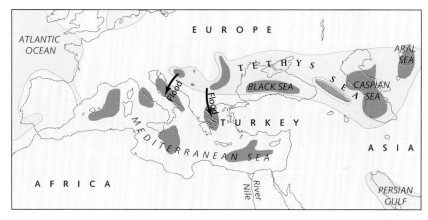

5 to 6.5 million years ago
A drop in sea levels left the Mediterranean region repeatedly isolated, and periodically reconnected to the Atlantic via south-west France. Freshwater animals are found in evaporite deposits from this time, suggesting that small lakes in the northern Mediterranean area were filled from the freshwater Tethys remnant covering the Black, Caspian and Aral Seas.

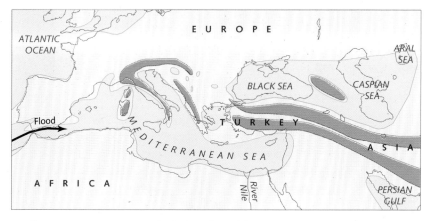

4 million years ago
Rising sea levels in the Pliocene period and the break through of the Straits of Gibraltar allowed the Atlantic to fill the Mediterranean basin. The Black Sea was joined to the Caspian and Aral Seas at this time though they have been separated for about the last 3 million years.

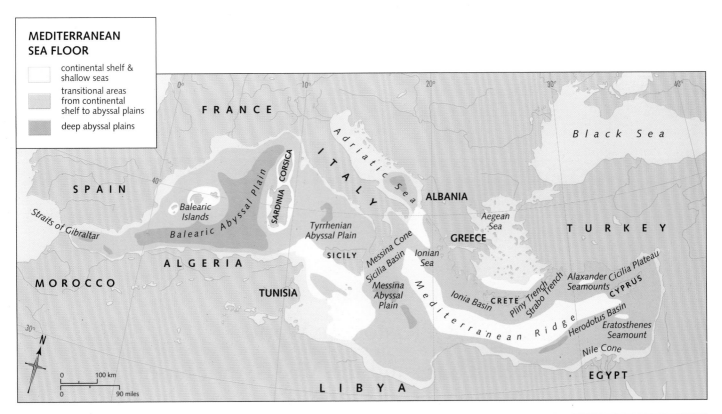

MEDITERRANEAN SEA FLOOR

- continental shelf & shallow seas
- transitional areas from continental shelf to abyssal plains
- deep abyssal plains

The Mediterranean sea floor

The topography of the sea floor of the Mediterranean shows great variation, reflecting its origins in the Alpine mountain-building episode. The Mediterranean is also the largest remaining remnant of the Tethys Sea. This body of water separated Africa and southern Europe from northern Europe, and Laurasia from Gondwanaland, for much of the Mesozoic era. It was at times a true ocean, with oceanic crust forming its sea-floor, and at other times a continental shelf sea. The marine fauna of the Tethys is found in fossil beds over a large area of Europe and the Middle East, making it one of the most studied parts of the Earth's history. The Balearic abyssal plain to the west and the Mediterranean Ridge to the east are separated by a shallow region stretching from Tunisia to Sicily. Until recently this was a land bridge allowing migrations of animals from Africa to Europe and vice-versa.

The Nile delta

One of the world's great rivers, the Nile empties its vast load of sand and silt into the Mediterranean. Being almost an inland sea, the Mediterranean has small tides, and less severe currents than open oceans, so there is no mechanism for washing away the sediment from the mouth of the Nile. It therefore accumulates, choking the mouth of the river and forcing it to disperse into small streams to find its way to the sea. An immense pile of sediment is pushed out into the Mediterranean by the force of the material behind it, forming the Nile cone.

Deltas are of enormous importance in the geological history of the Earth because of the vast amounts of material they accumulate in a relatively short time. Large parts of the continent of North America consist of ancient deltaic deposits, as do the Cleveland Hills of northern England, much of the North Sea basin and the Donets Basin in Russia.

Landsat image of the Nile delta (below)

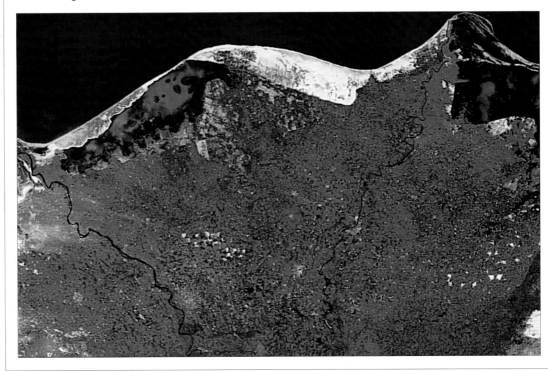

Part Seven
Ice Ages and the First Humans

Ice Ages

The Pleistocene World

The Great Lakes and the Mississippi Basin

Human Origins and Migration

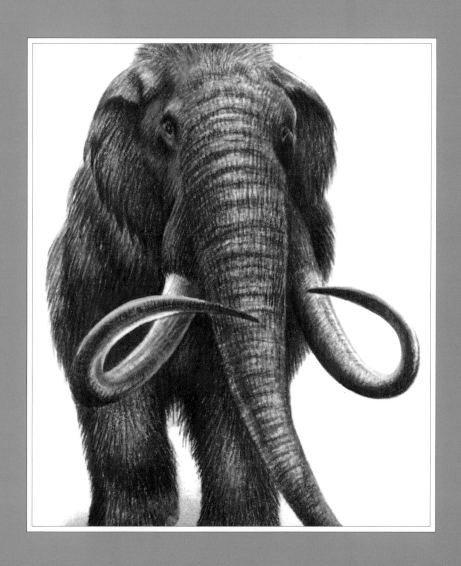

Ice Ages
The world turns colder

About 3 million years ago the Earth began to get colder. Throughout its history the average temperature on Earth has fluctuated, though it has remained within a surprisingly narrow range. On the geological time-scale we are in an exceptionally cold period, though warmer periods occur at regular intervals. At the present time we happen to be in an interglacial period – a warm spot between glaciations.

We know the reasons for the regular 10,000-year fluctuations in temperature that have brought about the advance and retreat of the ice throughout the last 2 million years. It is less easy to explain why the Earth is in its present long cold phase. Many scientists now believe the answer partly lies with the geography of the Earth, and in particular the positions of the continents in the polar regions.

The Earth's weather systems are influenced by the arrangements of the continents. During the early Mesozoic era, for instance, the continents were grouped together leaving vast areas of open deep ocean. This allowed surface currents to bring warm water from the equator to the poles, and cold water in the other direction. This created a smaller difference in climate between the poles and the tropics, and there were no ice caps for millions of years.

But now the continents are arranged so that the Antarctic is cut off from warm air and water by a circular current, and the Arctic is surrounded on all sides by continents. The beginning of this configuration coincided with, and may possibly have caused, the present cooling of the Earth.

The Milankovitch cycle
In 1920 a Yugoslavian meteorologist, Milutin Milankovitch made measurements of the changes in the levels of the Sun's radiation reaching the Earth. He was able to show that the heat from the Sun, and therefore the temperature of the Earth, is affected by three factors which occur in cycles: variations in the Earth's orbit around the Sun; variations in the tilt of the Earth's axis; and wobbles of the Earth's axis caused by gravitational interactions between the Sun, the Moon and the Earth. The cycles are 100,000, 41,000 and 22,000 years long respectively. These periods tie in well with the duration of recent ice ages.

Since glacial episodes are very uncommon in the Earth's history, it is generally agreed that the Milankovitch cycle has to be combined with other factors for a glacial episode to occur.

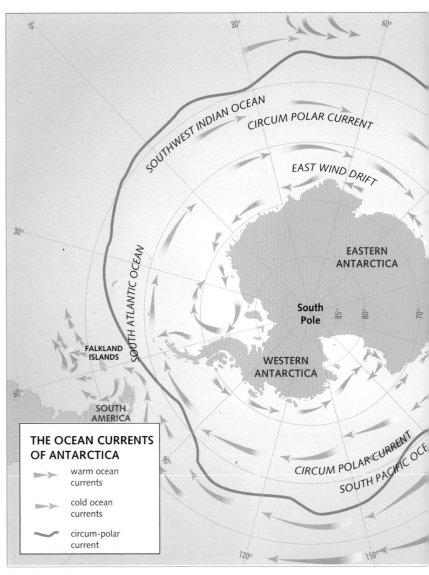

THE OCEAN CURRENTS OF ANTARCTICA

→ warm ocean currents

→ cold ocean currents

━ circum-polar current

The cold south
The continent of Antarctica severed its links with the other southern continents over the past 100 million years. This led to the formation of the Circum-Antarctic current, which effectively cut off the flow of warm water, and therefore warm air to Antarctica and to the polar region.

The cooling of the south polar region led to a drop in overall sea temperature, and the formation of an ice-cap caused further cooling of the Earth's atmosphere – an ice-cap reflects sunlight back without absorption. The drop in temperature may then have been enough to create an ice-cap at the North Pole, which in turn led to more cooling.

Global temperatures
Changes in the Earth's surface temperature have been irregular and largely unexplained.

Average annual temperature difference between a normal climate and a glacial one is less than 10°C. A drop of only 4–5°C in average annual temperature from the

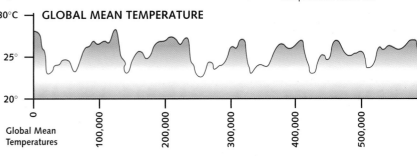

GLOBAL MEAN TEMPERATURE

30°C

25°

20°

0   100,000   200,000   300,000   400,000   500,000

Global Mean Temperatures

0   1280 km
0   800 miles

THE MOVEMENT OF ANTARCTICA

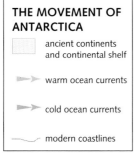

ancient continents and continental shelf

warm ocean currents

cold ocean currents

modern coastlines

Antarctica 60–62 million years ago

At the beginning of the Tertiary period South America and Antarctica were joined via the Antarctic peninsula and Cape Horn. Westerly ocean currents were forced north by this land barrier, allowing warm water to be carried south by the Brazil current. This inflow warmed the polar region, preventing the formation of an ice cap.

Antarctica 35 million years ago

Sea floor spreading began in this region at about this time, possibly caused by ocean ridges in the Pacific being subducted beneath the west coast of South America as it drifted to the west. As the continents began to separate there was still enough continental shelf to prevent the flow of the cold westerly current through the gap.

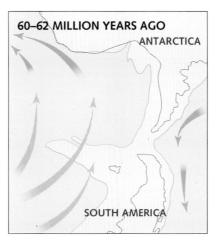

60–62 MILLION YEARS AGO
ANTARCTICA

SOUTH AMERICA

35 MILLION YEARS AGO
ANTARCTICA

SOUTH AMERICA

10 MILLION YEARS AGO
ANTARCTICA

SOUTH AMERICA

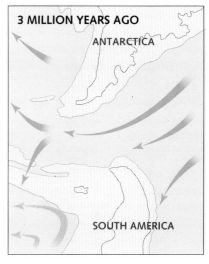

3 MILLION YEARS AGO
ANTARCTICA

SOUTH AMERICA

Antarctica 10 million years ago

A deep ocean lay between the continents and the Westerly Drift (the ocean current that flows round the Antarctic) could force its way through. It also pushed north, preventing the Brazil current from reaching the continent of Antarctica.

Antarctica 3 million years ago

The circum-Antarctic Current, or Westerly Drift, was now fully established. Warm water was turned back by the cold current, isolating the Antarctic region in a cold climate zone. Sea floor spreading had moved to the eastern edge of the Scotia island arc, which connects South America and Antarctica.

present (already slightly colder than normal) could cause renewal of continental glaciation.

Elephant Island, Drake passage *(right)*

The earth became dramatically colder over the last three million years. This coincided with the opening

of the sea between South America and Antarctica at Drake passage, allowing the circum-Antarctic current to develop.

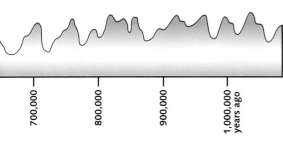

700,000

800,000

900,000

1,000,000 years ago

The Pleistocene World

The Earth 2 million years ago

For most of the last two million years the Earth has been unusually cold. A succession of 'ice ages' dominated the Pleistocene period, and profoundly affected the climate, geography and plant and animal life of the planet. Each glaciation lasted for about 100,000 years with warmer interglacial periods lasting for about 10,000 years in between. It is now about 8,000 years since the end of the last major ice age.

At the peak of the last ice age, 18,000 years ago, 5 per cent of the world's atmospheric and sea water was frozen in the glaciers and ice-caps. This caused sea levels to drop 100 metres (330 ft) below present levels. This drop in sea levels throughout the Pleistocene ice ages created land bridges that have had a major effect on the distribution of animal and plant life in our present world. The most important links include the Bering bridge from Siberia to Alaska, the North Sea bridge from Europe to the British Isles, the Sunda bridge between western Indonesia and Asia and the New Guinea to Australia bridge. Animals found that they could make their way across these bridges, often bringing spores and seeds of plants with them. The Bering land bridge is also thought to be the route by which humans first came to North America. Climate affected the movements across land bridges: Ireland, for example, has no reptiles, since its land bridge to Britain would only have existed when the temperature in the region was very cold, and unfavourable to reptiles.

In many ways, the Earth and its life forms are still responding to the return to warmer conditions and the retreat of the ice-sheets. In geological and evolutionary forms the time for adjustment has been very short. Some northern continents are still rising as the weight of the sheet has been removed, and many warm-climate plants and animals are moving north – back to their pre-ice age latitudes.

Western Canada
The Canadian Rockies were the centre of the Cordilleran ice-sheet which covered the eastern half of Canada and Alaska.

North America
The Laurentide ice sheet radiated out from Hudson Bay. It reached to the Arctic, covering eastern Canada, New England, and the northern midwest states.

Himalayas
Ice-sheets extended outwards from the Himalayas and from those parts of Siberia where the climate was wet enough to allow ice to form.

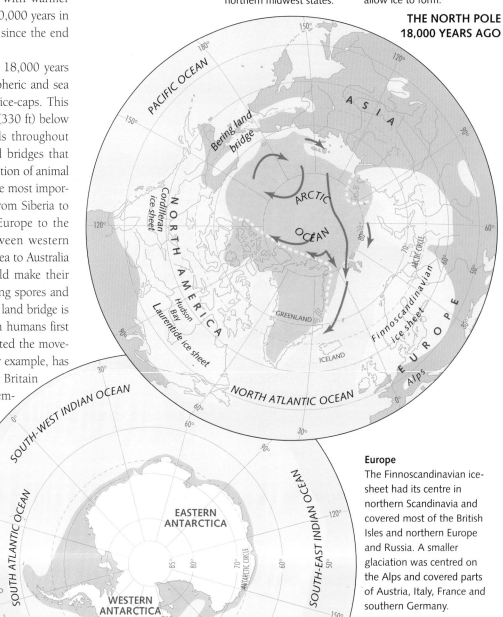

**THE NORTH POLE
18,000 YEARS AGO**

**THE SOUTH POLE
18,000 YEARS AGO**

Europe
The Finnoscandinavian ice-sheet had its centre in northern Scandinavia and covered most of the British Isles and northern Europe and Russia. A smaller glaciation was centred on the Alps and covered parts of Austria, Italy, France and southern Germany.

The Southern Hemisphere
The Antarctica ice-sheet extended to about double its present extent. Elsewhere small ice-sheets formed and expanded in the Andes and the mountains of Australia and New Zealand.

☐ ice cap 18,000 years ago

▨ present ice-cap

➜ direction of ice flow

The peak of the last ice age, 18,000 years ago

Each glaciation causes such alteration to the face of the Earth that it obliterates much of the evidence of previous ice ages. We know most about the last ice age, which reached its peak 18,000 years ago. At this time parts of the ice-sheets covering Canada, Greenland and northern Europe reached 3,000 metres (10,000 ft) in thickness. The world's average temperature was about 6°C lower than at present.

The build-up of ice depended on different factors, as it does in today's world. Ice-sheets form close to the poles and in high latitudes, where the temperatures are low all the year round. Glaciers also form at high altitudes, so mountain ranges acted as centres of glaciation. Glaciation needs a supply of water and this is normally brought in by movements of moist air. In parts of Siberia, for instance, temperatures were extremely cold throughout the Pleistocene, but little ice was formed because of the dryness of the atmosphere in the region.

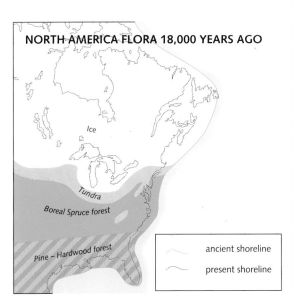

NORTH AMERICA FLORA 18,000 YEARS AGO

Ice

Tundra

Boreal Spruce forest

Pine – Hardwood forest

~~~ ancient shoreline
~~~ present shoreline

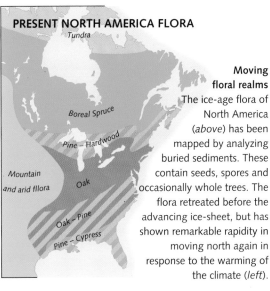

PRESENT NORTH AMERICA FLORA

Tundra

Boreal Spruce

Pine – Hardwood

Mountain and arid fllora

Oak

Oak – Pine

Pine – Cypress

Moving floral realms

The ice-age flora of North America (*above*) has been mapped by analyzing buried sediments. These contain seeds, spores and occasionally whole trees. The flora retreated before the advancing ice-sheet, but has shown remarkable rapidity in moving north again in response to the warming of the climate (*left*).

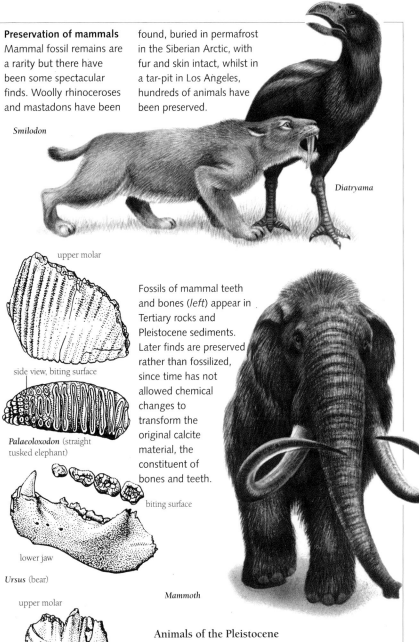

Preservation of mammals
Mammal fossil remains are a rarity but there have been some spectacular finds. Woolly rhinoceroses and mastadons have been found, buried in permafrost in the Siberian Arctic, with fur and skin intact, whilst in a tar-pit in Los Angeles, hundreds of animals have been preserved.

Smilodon

Diatryama

upper molar

side view, biting surface

Palaeoloxodon (straight tusked elephant)

Fossils of mammal teeth and bones (*left*) appear in Tertiary rocks and Pleistocene sediments. Later finds are preserved rather than fossilized, since time has not allowed chemical changes to transform the original calcite material, the constituent of bones and teeth.

biting surface

lower jaw

Ursus (bear)

Mammoth

upper molar

side view

biting surface

Anancus (mastadon)

Crocuta (cave hyaena)

biting surface

lower jaw

Animals of the Pleistocene

Faced with colder conditions, animals either adapted or moved south. Some, like the tigers did both, either following the grazing animals south from their Siberian origins, or adapting to colder conditions and different prey. Those mammals that lived on the fringes of the ice-sheets grew thick coats. Woolly mammoths and rhinoceroses are still being found buried in permafrost in Siberia. The migration of animals northwards continues today as armadillos appear in Florida, and opossum and raccoon in the Great Lakes area.

Sea levels affected marine life, in particular corals which lived close to the shore lines. In the West Indies, eroded platforms can be still be seen on reefs which were exposed to wave action by the drop in sea level. The reefs grew again as the ice melted and sea levels rose.

The Great Lakes and the Mississippi Basin

Remnants of the ice ages

At its northern edge the Mississippi Basin is moving northwards as the rivers erode the glacial deposits of the plains of the Midwest. The Illinois River is now within a few kilometres of Lake Michigan. The Great Lakes presently drain into the Atlantic via the St. Lawrence River but when the Illinois River reaches Lake Michigan this pattern will rapidly change. The Mississippi drainage system will then capture the Great Lakes and drain them into the Gulf of Mexico. This will be the climax of the natural process that has been going on since the end of the last ice age, as thousands of glacial remants, like the Great Lakes, have drained into ever-growing river basins.

In the last Ice Age North America experienced four separate glacial episodes, each interrupted by a slight warming and retreat of the edge of the ice sheet. Large areas of what is now the United States were covered in thousands of lakes. These appeared in all the regions the ice-sheet had reached: around the Great Lakes and north-east, and in the uplands of the west. The glaciers and ice-sheets had gouged small narrow rivers valleys into deep, broad depressions. When the ice retreated huge amounts of eroded material were dumped by meltwater, which often flooded areas stretching well beyond the extent of the original ice-sheets. This sediment blocked normal drainage channels, creating lakes and diverting rivers. Since then river drainage systems have developed which have eroded much of the glacial mud and drained many of the glacial lakes. These river systems, such as the Mississippi, are continuing to work their way towards the lakes that lie within their drainage basins.

Muldrow Glacier
The Strait of the Muldrow Glacier, part of the Mt. McKinley Massif, Alaska. The characteristic curve of glaciers is reflected in glacial remnant lakes.

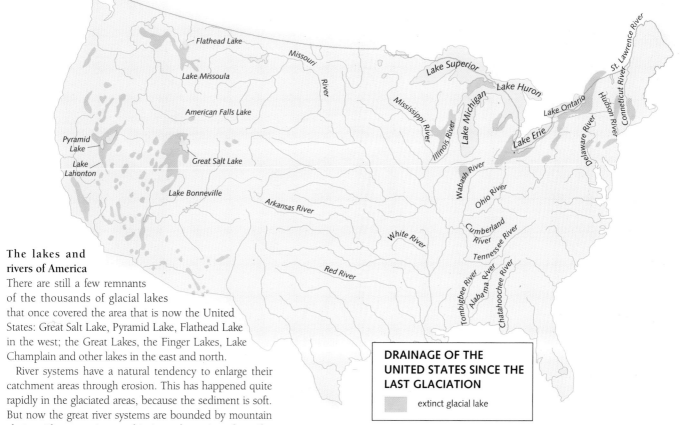

DRAINAGE OF THE UNITED STATES SINCE THE LAST GLACIATION
▢ extinct glacial lake

The lakes and rivers of America
There are still a few remnants of the thousands of glacial lakes that once covered the area that is now the United States: Great Salt Lake, Pyramid Lake, Flathead Lake in the west; the Great Lakes, the Finger Lakes, Lake Champlain and other lakes in the east and north.

River systems have a natural tendency to enlarge their catchment areas through erosion. This has happened quite rapidly in the glaciated areas, because the sediment is soft. But now the great river systems are bounded by mountain chains. The exception to this is at the point where the Illinois River, part of the Mississippi River system, almost meets Lake Michigan. The river will take a few thousand years to reach the lake, and sveral thousand more to erode a channel to the depth of the lake and, thereby drain it. But it is more likely than not that this will happen – unless another glaciation intervenes. The position of the Illinois River and its tributaries is clearly shown on the map (*top right*). When the Illinois River reaches Lake Michigan the Mississippi river system will carry the lake-water south to the Gulf of Mexico.

Niagara
The Niagara Falls lie on the Niagara River between Lake Erie and Lake Ontario. Over the next few thousand years it is likely that changes in drainage patterns will radically alter the flow of the rivers around the Great Lakes.

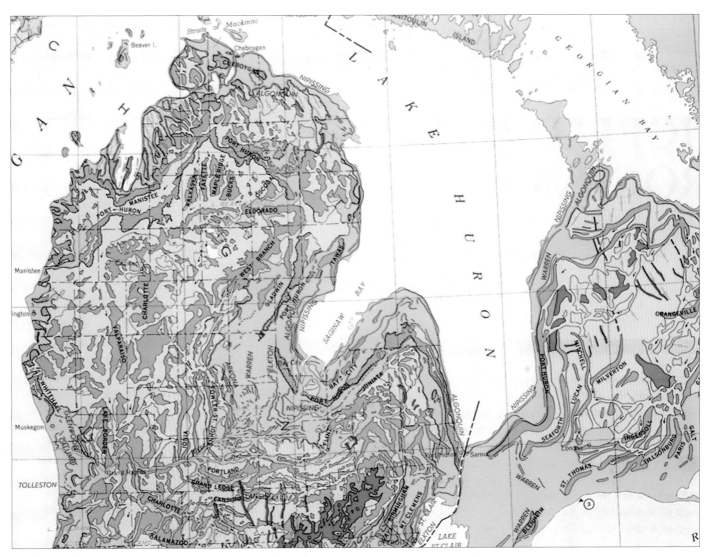

Landsat image of the Great Lakes (*right*)

The curved southern ends of the lakes reflect the shape of the ice-sheets at their southern limit. The region around the Great Lakes is extremely flat. In the area to the north the ice-sheets eroded most of the younger sediment, exposing the ancient Precambrian Canadian underneath, with its even topography. To the west and south glacial and river deposits have given an even covering of rich soil, that forms the Great Plains – one of the most fertile areas in the world.

Glacial features of the southern Great Lakes (*above*)

The effects of the two most recent ice-sheets can be clearly seen on this map of glacial sediments around one portion of the Great Lakes. The Illinoian glaciation pushed the edge of the ice-sheet over 500 kilometres (300 miles) south of the Great Lakes. When it retreated it melted and left a broad band of sediment that had been carried south in the frozen sheet. Later came the Wisconsin glaciation. This reached 200 kilometres (125 miles) south and then retreated in stages. The last ice-sheet disappeared from the south of the Great Lakes 13,000 years ago. By 9,000 years ago, the edge of the ice had retreated to the north shore of Lake Superior, and by 4,000 years ago it was confined to northern Canada, Baffin Island and Greenland. Deposits, known as terminal moraines, were made at the edges of the ice-sheet. These echo the shapes of the Great Lakes, which were formed by ice-sheets gouging out depressions in the ground.

Lake sediments indicate the positions of the thousands of lakes that were created by the retreating ice-sheets. These have been drained by rivers. The Great Lakes themselves were much larger than at present.

Human Origins and Migration
From Africa to every corner of the Earth

In all of human science nothing has caused more dispute than the origins of our own species. We do know that remains of our earliest direct ancestors have been found only in Africa. From this evidence it seems that Homo erectus, our most likely ancestor, walked out of Africa about 1.5 million years ago. Homo sapiens, our own species, evolved much later – about 250,000 years ago – again probably in East Africa. This new hominid then migrated across the world and replaced those related species that had preceded it.

The technological sophistication of *Homo sapiens*, the new species, enabled them to adapt easily to different conditions. They arrived in east Asia within 40,000 years and into Australasia soon afterwards. They would have used the land bridges that were available in the ice ages, but there is some evidence that they used boats as well. Homo sapiens probably did not arrive in Europe until 40,000 years ago. The migration into North America is even more recent.

Around 30,000 to 20,000 years ago humans followed the migration route of mammoths, reindeer and other mammals across the Bering Land Bridge from Siberia to Alaska. This may have happened as sea levels dropped during the Wisconsian glaciation. After being isolated in Alaska for several thousand years humans moved south as a corridor opened up in the ice-sheets between the Rocky Mountains of the Laurentian glacial centre.

The earliest definite dates of habitation in North America are 12,000 years ago for Stone Age peoples of the southwestern United States. Humans had reached the southern end of South America, only 2,000 years later. Most parts of the Americas were inhabited by humans 10,000 years ago.

Australopithecus

The earliest hominids are named *Australopithecus* which means literally 'southern ape'. Remains have been found in Ethiopia and East and South Africa. Because finds are rare, there is much disagreement over the interpretation of the fossils.

Australopithecus afarensis is the earliest known hominid; remains from Ethiopia date back 3.5 million years. *Australopithecus africanus*, which lived 3 to 1 million years ago, had limbs and teeth similar to modern man. *Australopithecus robustus* lived at the same time as *A. africanus*, but whether they share a common ancestor is uncertain. The origin of *Australopithecus* itself is unknown. A species called *Sivapithecus* shared characteristics of primates and hominids, but its status as a missing link is undermined as it died out well before *Australopithecus* appeared.

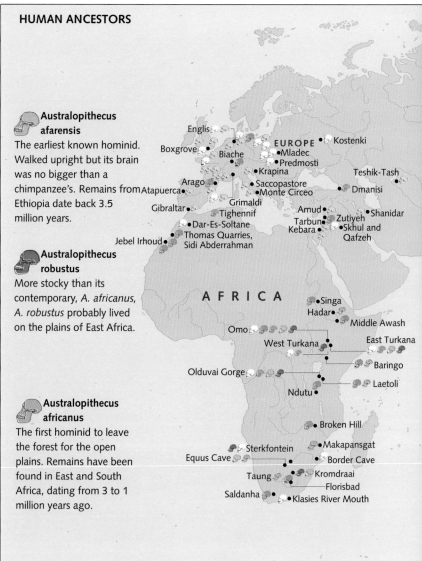

HUMAN ANCESTORS

Australopithecus afarensis
The earliest known hominid. Walked upright but its brain was no bigger than a chimpanzee's. Remains from Ethiopia date back 3.5 million years.

Australopithecus robustus
More stocky than its contemporary, A. africanus, A. robustus probably lived on the plains of East Africa.

Australopithecus africanus
The first hominid to leave the forest for the open plains. Remains have been found in East and South Africa, dating from 3 to 1 million years ago.

EUROPE
Englis
Boxgrove
Biache
Kostenki
Mladec
Predmosti
Krapina
Teshik-Tash
Arago
Saccopastore
Atapuerca
Monte Circeo
Dmanisi
Gibraltar
Grimaldi
Amud
Shanidar
Tighennif
Tarbun
Zutiyeh
Dar-Es-Soltane
Kebara
Skhul and
Jebel Irhoud
Thomas Quarries,
Qafzeh
Sidi Abderrahman

AFRICA

Singa
Hadar
Middle Awash
Omo
West Turkana
East Turkana
Baringo
Olduvai Gorge
Laetoli
Ndutu

Broken Hill
Sterkfontein
Makapansgat
Equus Cave
Border Cave
Taung
Kromdraai
Florisbad
Saldanha
Klasies River Mouth

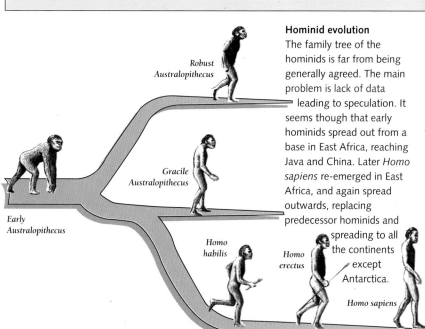

Robust Australopithecus

Gracile Australopithecus

Early Australopithecus

Homo habilis

Homo erectus

Homo sapiens

Hominid evolution
The family tree of the hominids is far from being generally agreed. The main problem is lack of data leading to speculation. It seems though that early hominids spread out from a base in East Africa, reaching Java and China. Later *Homo sapiens* re-emerged in East Africa, and again spread outwards, replacing predecessor hominids and spreading to all the continents except Antarctica.

Homo erectus
The first real human emerged 1.5 million years ago. Remains have been found in Java, China and the Caucasus.

Homo heidelbergenis
A likely descendant of Homo erectus. Remains have been found in both Europe and China.

Homo neanderthalensis
Remains of this very close relative of modern humans have been found all over Europe. Neanderthals died out 35,000 years ago.

Homo sapiens
Our immediate ancestor emerged in Africa about 250,000 years ago. It quickly spread out of Africa, replacing the remaining 'Homo'.

Homo

The term 'Homo' is used for creatures belonging to the same group as modern humans. Brain size is a guide to membership of this group. *Homo habilis* is the earliest *Homo*, with a brain about half the size of a modern human's, but 50 per cent bigger than any *Australopithecus*. It emerged about 2 million years ago in East and southern Africa. Its feet and hands show similarities to modern humans, and some tools have been found near the remains. *Homo erectus* emerged about 1.5 million years ago. It had the ability to use fire for warmth, cooking and hunting, and was the first hominid human to leave Africa. *Homo neanderthalensis* lived at the same time as our immediate ancestor *Homo sapiens*, and there may have been some interbreeding before the former died out 35,000 years ago.

Homo habilis (handy man)
Fossils indicated hands with a strong grip and that the creature could walk and run.

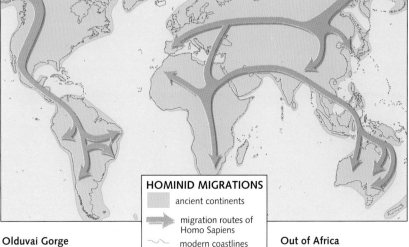

HOMINID MIGRATIONS
ancient continents
migration routes of Homo Sapiens
modern coastlines

Olduvai Gorge
The fossils found at this famous site in East Africa include a hominid with a much larger brain than Australopithecus, Homo habilis. Remains of

Australopithecus have been found alongside *Homo habilis* calling into question the idea that the former is the ancestor of the latter.

Out of Africa
Homo erectus migrated from Africa and reached sites as far away as Java and China. Homo sapiens followed several million years later.

Part Eight
Present and Future Earth

Mount St. Helens

The San Andreas Fault

Earthquakes

Antarctica

The Earth's Climate

Future World

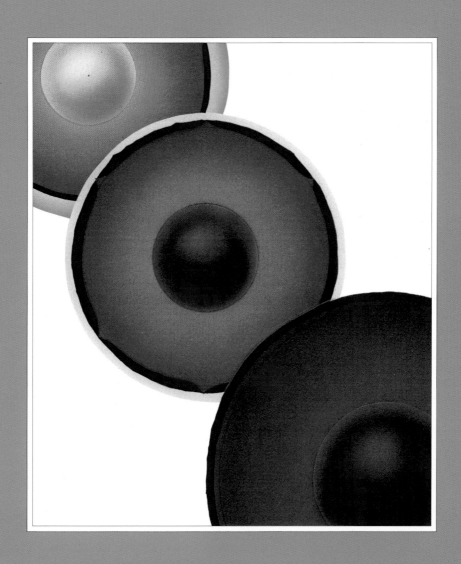

Mount St.Helens

Volcanoes in recent history

Mount St. Helens is one of a row of towering peaks in the Cascade Range that runs from northern California to southern Canada. On March 20th, 1980 seismologists picked up signs of earthquake activity below the mountain.

In April 1980 earth tremors began to indicate that large amounts of magma were moving beneath the mountain. At 8.32 a.m. on May 18th a large earthquake 3 kilometres (2 miles) below the summit triggered a massive landslip on the northern side of the mountain, exposing the reservoir of magma and superheated water. This was immediately followed by a jet of superheated ash, gas and steam exploding out of the side of the mountain with immense force. A surge of airborne magma and debris then blew the whole side of the mountain away. The energy of the eruption was equivalent to a hydrogen bomb of about 25 megatons, 1,300 times as powerful as the bomb that destroyed Hiroshima. Sixty people were killed, and weather patterns around the world were disrupted.

Despite the significance of the eruption in human terms, seen in a geological context it was a minor event. The calderas at Yellowstone and Long Valley have shown recent signs of resurgence – if either were to erupt on a similar scale to their past explosions, an area of some 30,000 square kilometres (12,000 square miles) would be covered in ignimbrite (rock formed from volcanic ash) with almost total loss of life. The blocking of sunlight would disrupt climate and agriculture all over the world. Eruptions of this magnitude have occurred perhaps ten times in the past million years.

Mount St. Helens
Volcanic ash rises 18 km (5 miles) from the eruption of Mount.St. Helens. The major eruption blew away 3 cubic km (.72 cubic miles) of the mountain and devastated more than 500 km (310 miles) of land.

The eruption of Mount St.Helens

When, on the morning of May 18th, a jet of ash, gas and steam exploded out of the mountain, it blew over vehicles and melted plastic seats 26 km (16 miles) away, and devasated an area 20 km (12 miles) out from the volcano and 30 km (19 miles) wide. A vertical eruption sent an ash plume to a height of 25 km (15 miles). The ash cloud drifted to the east and north-east on prevailing winds and brought darkness at noon to an area 250 km (155 miles) to the east.

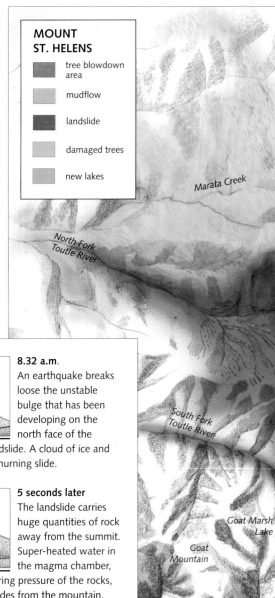

MOUNT ST. HELENS

- tree blowdown area
- mudflow
- landslide
- damaged trees
- new lakes

Marata Creek

North Fork Toutle River

South Fork Toutle River

Goat Marsh Lake

Goat Mountain

Kalama River

8.32 a.m.
An earthquake breaks loose the unstable bulge that has been developing on the north face of the mountain. This starts a landslide. A cloud of ice and rock is thrown up by the churning slide.

5 seconds later
The landslide carries huge quantities of rock away from the summit. Super-heated water in the magma chamber, kept liquid by the overbearing pressure of the rocks, flashes to steam and explodes from the mountain.

15 seconds later
As the magma chamber is uncovered melted ice and groundwater is heated by contact with the magma. The magma explodes from the north face, sending super-heated ash and debris in a huge cloud.

25 seconds later
The north face of the mountain has been blown away. Clouds of ash, steam and volcanic debris pour out. Magma swells up and runs out of the north face. A crater 3 km (2 miles) wide is left by the eruption.

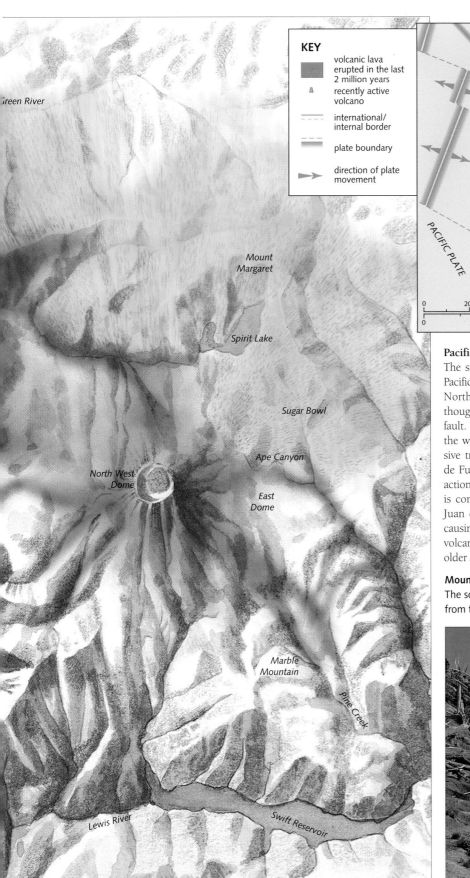

KEY

- ▨ volcanic lava erupted in the last 2 million years
- ⌂ recently active volcano
- – – – international/internal border
- ▬▬▬ plate boundary
- ➤ direction of plate movement

Pacificic plates

The small Juan de Fuca plate is sandwiched between the Pacific and North American plates. The western edge of the North American plate runs under the sea at this point, though it runs on land further south as the San Andreas fault. The Pacific Ridge was fractured by its collision with the westward-moving North American plate, causing massive transform faults that run under the Pacific. The Juan de Fuca plate is being pushed eastwards by the spreading action of the Pacific Ridge, while the North American plate is continuing to move westwards. The result is that the Juan de Fuca is being subducted under North America, causing volcanic activity in the Cascade range. The Cascade volcanoes represent a new volcanic arc superimposed on older structures.

Mount St. Helens

The scene of devastation to forest and land caused by lava from the Mount St. Helens eruption.

The San Andreas Fault

Plate margins in the present world

The San Andreas fault is one of the most famous geological features in the world – and with some justification. For this is the only place on Earth where the boundary between two of the crust's plates passes through such heavily populated areas. The San Andreas fault is the point where the edge of the North American continent rubs up against the plate that makes up the Pacific Ocean floor. Normally this boundary would lie off shore at the margin of the continental shelf. But the movements of the plates over the last 100 million years have produced the situation we see today, where a sliver of continent is being hauled up the west coast of America, rubbing along the edge of the plate as it goes. The plates move relative to each other at a rate of about 1 cm (.4 in) per year at present – though they do not move constantly, they tend to stick and jerk or snap, and it is this that causes earthquakes. The piece of continent to the west of the fault is estimated to have moved north by over 500 kilometres (310 miles) in the past 100 million years.

Present geological disturbances in western America can be traced back to the collision between westward-moving North America and part of the Pacific Ridge system during the Cenozoic era. This continent-ridge collision created huge transform faults. Movement along these faults has since offset large segments of the ridge, the sea-floor and even the edge of the continent itself. The San Andreas fault is a great transform fault which has offset the Pacific Ridge 3,000 km (2,000 miles) and sliced off a long sliver of continental crust. Volcanic arc activity continues in those places where the continent was still to the west of the ridge, and therefore ocean crust was still being subducted.

Northern California collided with the Pacific ridge about 30 million years ago. As North America continued to move west, more and more of the continental margin hit the ridge, and the San Andreas fault grew in length. Baja California was torn away from the mainland 5 million years ago, and the gulf of California began to form. The sliver of continent now lying west of the fault was detached from the North American plate and has since moved as part of the Pacific plate. Eventually western California will become either completely detached as a microcontinent or will find itself joining western Canada.

Earthquakes in California

The San Andreas fault has what is known as a shallow depth (c.10 km) of rupture. This gives rapid damping of seismic waves in California which is reassuring for those who do not live in the immediate vicinity of the fault. However, there are several pockets of high intensity that are obvious danger areas. Towns along the east side of the fault (on the American plate) are less vulnerable to damage.

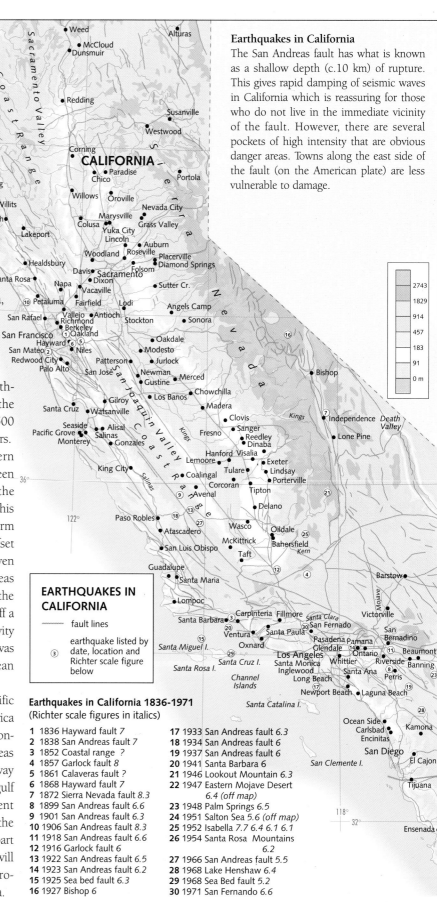

EARTHQUAKES IN CALIFORNIA

— fault lines

③ earthquake listed by date, location and Richter scale figure below

Earthquakes in California 1836-1971
(Richter scale figures in italics)

1	1836 Hayward fault *7*	17	1933 San Andreas fault *6.3*
2	1838 San Andreas fault *7*	18	1934 San Andreas fault *6*
3	1852 Coastal range *?*	19	1937 San Andreas fault *6*
4	1857 Garlock fault *8*	20	1941 Santa Barbara *6*
5	1861 Calaveras fault *?*	21	1946 Lookout Mountain *6.3*
6	1868 Hayward fault *7*	22	1947 Eastern Mojave Desert
7	1872 Sierra Nevada fault *8.3*		*6.4 (off map)*
8	1899 San Andreas fault *6.6*	23	1948 Palm Springs *6.5*
9	1901 San Andreas fault *6.3*	24	1951 Salton Sea *5.6 (off map)*
10	1906 San Andreas fault *8.3*	25	1952 Isabella *7.7 6.4 6.1 6.1*
11	1918 San Andreas fault *6.6*	26	1954 Santa Rosa Mountains
12	1916 Garlock fault *6*		*6.2*
13	1922 San Andreas fault *6.5*	27	1966 San Andreas fault *5.5*
14	1923 San Andreas fault *6.2*	28	1968 Lake Henshaw *6.4*
15	1925 Sea bed fault *6.3*	29	1968 Sea Bed fault *5.2*
16	1927 Bishop *6*	30	1971 San Fernando *6.6*

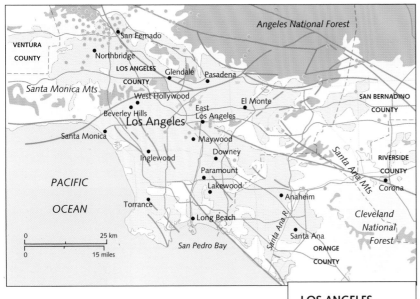

The San Francisco earthquake

On 18 April 1906, at 5.12 a.m. a section of rock a few kilometres from the Golden Gate bridge snapped along the San Andreas fault. The break spread north and south along the fault from Cape Mendocino. The subsequent earthquake registered 8.3 on the Richter Scale. A fire raged for three days, destroying most of the city's wooden buildings. The fire caused maybe ten times more damage than the earthquake. In total, 700 lives were lost.

The area affected by the earthquake was surprisingly small. The places of most severe rupture were limited to within a few tens of kilometres of the fault. The earthquake was felt as far north as Oregon and as far south as Los Angeles, a distance of 1,170 km. (720 miles).

The Los Angeles area

Smaller transform faults complicate the geological pattern in the Los Angeles area. The main San Andreas fault system has remained locked in the area north of Los Angeles since 1857. While south-east of Los Angeles there are frequent small earthquakes. The smaller earthquakes indicate movement along some of the 'swarm' of lesser faults and may be useful in relieving the stress built up by the movement of the plates. Heavily populated areas of greater Los Angeles lie within the range of the faults to the north of the city, which may be the most dangerous region, though few areas are free of risk.

LOS ANGELES

	urban area
	forest
	park
	late Cenozoic basins
	earthquakes
	fault lines
	main roads

San Andreas fault *(below)*

At this location the fault is clearly reflected in the topography. Elsewhere movement across the fault can be seen in roads and fences that no longer meet, or at the famous winery near Hollister, where some storehouses are built over the fault. But the San Andreas is more properly seen as a fault system, with movement taking place at a number of different locations along a complex system of parallel fault lines. This makes the location of any earthquake exteremly difficult to predict.

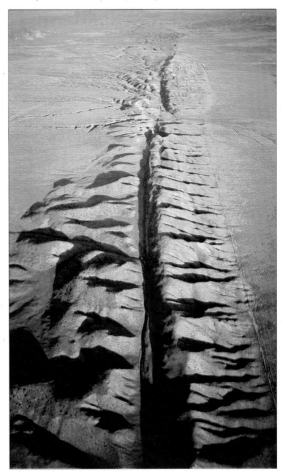

Earthquakes
Historical recordings of geological events

Earthquakes are part of the geological history of the Earth. They are also part of human history. More than 10,000 people die each year from the effects of earthquakes, and occasional catastropic earthquakes, which are frequently combined with volcanic eruptions, can alter the course of whole civilizations.

Earthquake centres often move around in discernible patterns, showing that plate boundaries are complex regions reflecting underlying movements in the Earth's mantle, rather than simple lines that can be drawn on a map. Written records pinpoint the locations and intensities of earthquakes much more exactly than geological evidence. In those parts of the world where historical records go back over thousands of years, we can learn something about changes in geological processes over comparatively short periods. Iran, for example, lies across the junction of four of the crust's plates, all moving relative to each other. The Iranian region is consequently crisscrossed by major faults and is subject to regular large-scale earthquakes. The region also has one of the Earth's longest recorded histories. It is close to the fertile crescent, the cradle of Near Eastern civilizations, and the region's position on major trading routes between the Mediterranean and India has also allowed others to record reports of earthquakes from the earliest times.

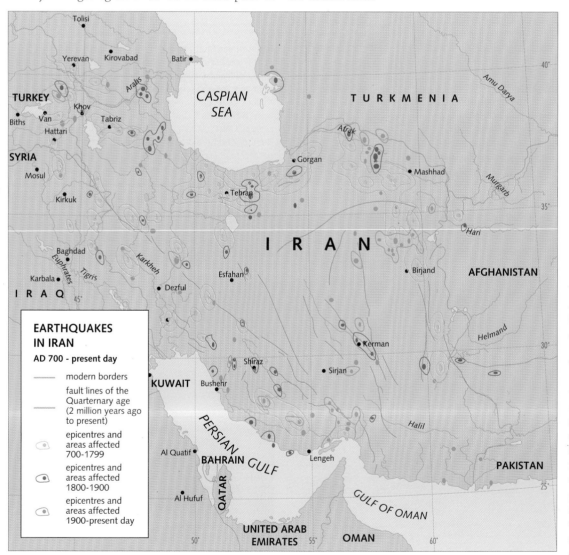

EARTHQUAKES IN IRAN

AD 700 - present day

— modern borders

— fault lines of the Quarternary age (2 million years ago to present)

epicentres and areas affected 700-1799

epicentres and areas affected 1800-1900

epicentres and areas affected 1900-present day

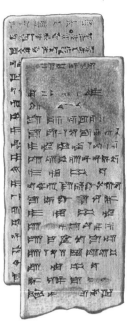

Early written evidence
The earliest earthquake to appear in written evidence in this region happened in about 300 BC. According to Poseidonius of Apameia, writing in around 250 BC, 'numerous cities and 2,000 villages were destroyed'. Later writers state that the city of Rhagae was destroyed and rebuilt by Seulucus Nector who died in 280 BC. But Alexander the Great was known to have passed through the city of Rhagae in 330 BC, before it was destroyed. This gives an approximate fix on the date of what was probably a series of major earthquakes. The illustration shows a letter from Ninenveh, written on a cuneiform tablet, which describes an earthquake in Assyria.

Earthquakes in Iran over 1,200 years

The complexity of the plate boundaries in the region around Iran is reflected in the locations of earthquakes, and in their movement. The movement shows some definite patterns – in some areas earthquakes have occurred in the 20th century and in the years before 1800, but not in the 19th century. The emergence of these patterns over such short time scales reflects underlying disturbances. Over geological time, plate boundaries are plotted as simple approximations. Historical evidence shows they are often complex zones containing swarms of faults, with earthquakes switching back and forth between different areas.

Traditional yurt
Circular tents which have been used as dwellings by the nomads of Central Asia for thousands of years, are unaffected even by major earthquakes.

Spherical construction
Built in the early 1970s in northern Teheran as a method of resisting earthquake damage. More conventional methods include erecting buildings on concrete rafts.

EARTHQUAKES IN HUMAN HISTORY

position on map	date	location	number of casualties
1	856	Corinth, Greece	45,000
2	1038	Shansi, China	23,000
3	1057	Chihli, China	25,000
4	1170	Sicily	15,000
5	1290	Chihli, China	100,000
6	1293	Kamakura, Japan	30,000
7	1456	Naples, Italy	60,000
8	1531	Lisbon, Portugal	30,000
9	1667	Shemaka, Caucasia	80,000
10	1693	Catania, Italy	60,000
11	1693	Naples, Italy	93,000
12	1731	Peking, China	100,000
13	1737	Calcutta, India	300,000
14	1755	Northern Persia	40,000
15	1755	Lisbon, Portugal	30-60,000
16	1783	Calabria, Italy	50,000
17	1797	Quito, Ecuador	41,000
18	1822	Aleppo, Asia Minor	22,000
19	1828	Echigo, Japan	30,000
20	1868	Peru and Ecuador	25,000

position on map	date	location	number of casualties
21	1875	Venezuela and Columbia	16,000
22	1897	Assam, India	1,500
23	1898	Japan	22,000
24	1906	Valparaiso, Chile	1,500
25	1906	San Francisco, USA	700
26	1907	Kingston, Jamaica	1,400
27	1908	Messina, Italy	160,000
28	1915	Avezzano, Italy	30,000
29	1920	Kansu, China	180,000
30	1923	Tokyo, Japan	99,000
31	1930	Apennine Mountains, Italy	1,500
32	1932	Kansu, China	70,000
33	1935	Quetta, Baluchistan	60,000
34	1939	Chile	30,000
35	1939	Erzincan, Turkey	40,000
36	1948	Fukui, Japan	5,000
37	1949	Ecuador	6,000
38	1950	Assam, India	1,500
39	1954	Northern Algeria	1,500
40	1956	Kabul, Afghanistan	2,000

position on map	date	location	number of casualties
41	1957	Northern Iran	2,500
42	1960	Southern Chile	5,700
43	1960	Agadir, Morocco	12,000
44	1962	North-western Iran	12,000
45	1963	Skoje, Yugoslavia	1,000
46	1970	Peru	20,000
47	1972	Managua, Nicaragua	10,000
48	1976	Guatemala	23,000
49	1976	Philippines	3,100
50	1976	New Guinea	9,000
51	1976	Iran	5,000
52	1977	Romania	1,500
53	1978	Iran	15,000
54	1980	Algeria	3,500
55	1980	Italy	4,000
56	1981	Iran	3,000
57	1982	West Arabian Peninsula	2,800
58	1983	Turkey	1,300
59	1985	Mexico City	9,500
60	1995	Kobe, Japan	5,429

Antarctica
Ice movements on a continental scale

Antarctica can seem like a vast dead continent to humans – inhospitable and static, a bare rock landmass covered in a massively thick ice sheet. In fact, the Antarctic ice-sheet is complex and dynamic. It is also crucial to the world's climate. For that alone it deserves close study as well as for its value as the last great wilderness on Earth.

The continent of Antarctica has an area of 14 million square kilometres (5.4 million square miles) – one-third larger than Europe and close to the size of South America. The ice-sheet buries all of the topographic features of the continent, though the surface of the ice-sheet does reflect the landscape underneath. Thick ice-sheets form because there is virtually no melting from year to year. The mean monthly temperature at the South Pole in summer is –30°C, in winter it is –54°C. In some areas the temperature has dropped to –75°C.

Although the Earth's polar regions are always colder than the tropics because of their greater distance from the Sun, the difference is more extreme now than is normal in the Earth's history. The configuration of the continents has led to cooling at the poles and an area of high atmospheric pressure over Antarctica keeps the region cold and prevents precipitation. This situation could change very rapidly once warm air or water found its way into the refrigerator that currently exists around Antarctica.

For most of its existence Antarctica was joined to Africa, South America, India and Australia in the southern supercontinent of Gondwanaland. Gondwanaland remained largely in the southern hemisphere throughout the 500 million or so years of its existence. But during that time Antarctica moved away from the South Pole and Gondwanaland was free of ice-sheets for much of its history.

There is a great deal of evidence of previous warmer climates on the continent of Antarctica. The coal seams that run through the Transantarctic Mountains are part of one of the largest coalfields in the world. This is proof of lush vegetation supported by a warm climate. Discoveries of non-fossilized wood have shown that large forests grew on the flanks of the Transantarctic Mountains 4 million years ago. Marine fossils from warm seas have also been found.

Despite the difficulties of finding fossils, the history of Antarctica is being discovered. The presence of particular fossil flora or fauna in South America and Australia, has led geologists to assume their contemporaneous existence on Antarctica.

Ross Sea ice shelf
The inlets of the Ross Sea and Weddell Sea are covered in ice all year round. Ice flows bring sheets down from the high ground of the continent to sea level. This moves across the ice shelves, before calving off into ice bergs when it meets warmer water.

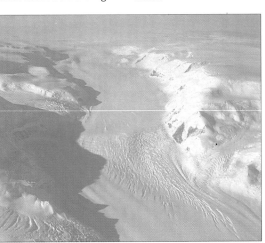

Giant glacier in Antarctica
The photograph (*above*) shows the Liv Glacier in the Queen Maud range of the Trans-Antarctic mountains, the dominant feature to the east of the Ross ice shelf

Antarctic Peninsula

Ellsworth Highland

BELLINGSHAUSEN
SEA

PACIFIC

AMUNDSEN
SEA

OCEAN

Thickness and volume of ice

The East Antarctic ice sheet covers 75 per cent of the area of the continent of Antarctica. It contains 80 per cent of Antarctica's ice by volume. The remainder is contained in the West Antarctic ice-sheet (14 per cent by area, 11 per cent by volume), the ice-shelves and the Antarctic peninsula. The ice-sheets have grown in thickness because, although there is very little precipitation (i.e. snow) in some areas, there is enough to compensate for the tiny amount of melting that takes place. This may be changing though, as the warming of the Earth since the last ice age now seems to be affecting Antarctica. Ships became stuck in ice in the Bellinghausen Sea in the early years of this century, in seasons which are now comparatively ice free; and plants that have never been seen on Antarctica are now beginning to establish themselves.

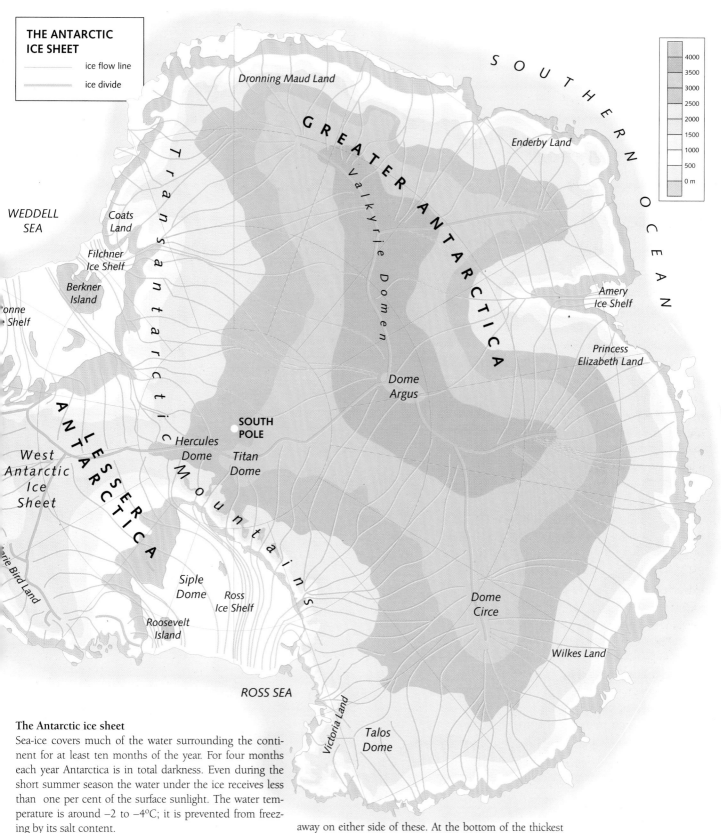

SOUTHERN OCEAN

Dronning Maud Land

GREATER ANTARCTICA

Valkyrie Domen

Enderby Land

WEDDELL
SEA

Coats
Land

Transantarctic

Filchner
Ice Shelf

Berkner
Island

onne
Shelf

Amery
Ice Shelf

Princess
Elizabeth Land

Dome
Argus

LESSER ANTARCTICA

West
Antarctic
Ice
Sheet

**SOUTH
POLE**

Hercules
Dome

Titan
Dome

Mountains

rie Bird Land

Siple
Dome

Ross
Ice Shelf

Dome
Circe

Roosevelt
Island

Wilkes Land

ROSS SEA

Victoria Land

Talos
Dome

4000
3500
3000
2500
2000
1500
1000
500
0 m

The Antarctic ice sheet

Sea-ice covers much of the water surrounding the continent for at least ten months of the year. For four months each year Antarctica is in total darkness. Even during the short summer season the water under the ice receives less than one per cent of the surface sunlight. The water temperature is around –2 to –4°C; it is prevented from freezing by its salt content.

The height of the ice surface above sea level is shown on this map. The ice flows away from its highest points towards the sea on all sides. As with river drainage, there are 'divides' which are the high ridges. The ice-sheet flows away on either side of these. At the bottom of the thickest ice-sheets plastic deformation takes place in the ice due to the pressure from above. This makes the ice flow more easily over the bedrock of the continent. The West Antarctic ice-sheet is thought to be unstable because of this.

The Earth's Climate
Present trends and future changes

We are now living in a warm period between ice ages. If there were no humans the Earth might now gradually become colder and the ice sheets begin to advance, although there are some other factors, besides human activity, which seem to be working against this. However, warmer conditions mean higher sea levels – and a high proportion of the world's human population live near sea level.

Weather forecasting can work with a reasonable degree of accuracy for about ten days into the future. After that the possible variations become impossible to predict. Changes in past climate are studied for clues as to how trends have developed, and how they might develop in the future. Unfortunately human activity in the past 200 years has affected the Earth's atmosphere in ways that are still poorly understood – the causes and even the existence of global warming are a matter of dispute.

In order to isolate natural processes we need to look further back, and it is then that geological evidence becomes important. Although geological events seem to have happened in an orderly fashion over many millions of years, conditions on the Earth can, in fact, change very rapidly. To understand how this might happen scientists concentrate on critical climatic systems. Study of the Antarctic ice cap is crucial to our understanding of present weather systems and possible future changes.

If, as many scientists suspect, the West Antarctic ice sheet does slide into the ocean it will profoundly alter the world's climate system in a relatively short time. The Circum-Antarctic current, the cold ocean current which encircles Antarctica, will be disturbed, and it seems likely that any diversion will allow some warm water to reach Antarctica. This will cause further warming, and therefore melting of the ice sheet. Reflection of the Sun's radiation by the ice -sheet will be reduced. Instead, underlying rocks will absorb warmth, causing more warming of the region. This is likely to change the Antarctic atmosphere from a cold high pressure zone to a low pressure zone. This could happen in a single year, and would immediately alter the global circulation patterns in the atmosphere which drive the world's weather..

The flooding of Europe and N. America *(above and right)*
Historical factors have led most of the world's population to make their homes within reach of the sea, or at low altitudes. This is where the most fertile soils are, and where trade is most easily carried out. But if the world's ice caps melt then the subsequent rise in sea levels will flood most of the world's major cities and areas of densest population.

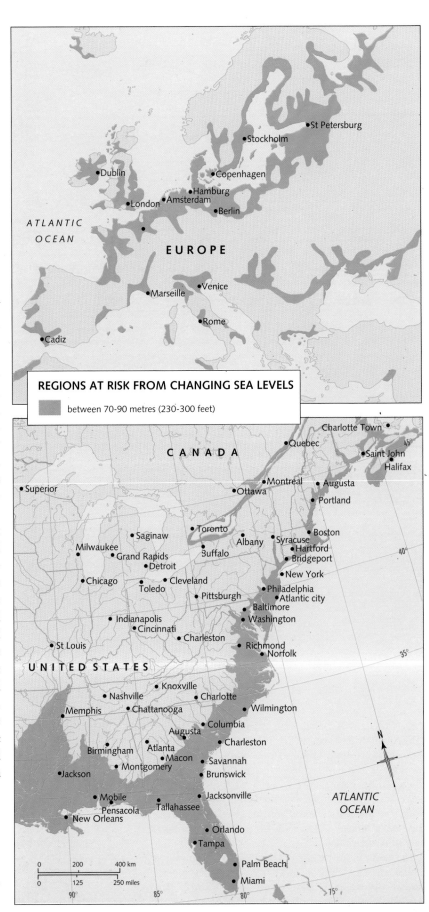

REGIONS AT RISK FROM CHANGING SEA LEVELS

between 70-90 metres (230-300 feet)

Sea level changes

Sea levels are dictated by two factors – the height at which continents float on the mantle and the amount of water in the world's oceans. The height of the continents can change in relatively short periods, usually because of melting ice caps relieving a huge burden of weight from them. But the effect of rising continents is countered by the increase in total sea water that occurs when ice-caps retreat. This can be due to overall warming or changes in local conditions in polar regions. Geological records have shown this can happen very rapidly. For example, ice cores from Greenland have shown that the overall temperature of the Earth dropped by 7°C in three years at the end of the last ice age.

An initial rise in sea levels would disturb ocean currents worldwide, making it likely that more warm water would penetrate the polar regions, causing more melting, and setting in chain its own cycle of melting and warming. The melting of the whole Antarctic ice cap would result in a rise in sea level of 70 metres. The melting of the Arctic would have less effect since most of the ice is floating, but melting of the Greenland ice cap alone would induce a rise of around 8 metres.

Cooling and warming in history

Within the overall warming of the planet since the end of the last ice age around 13,000 years ago, there have been cold episodes. An episode of cooling known as the Little Ice Age persisted in Europe from about 1450–1890, with particularly cold stages from 1600–1700 and 1800–1890. The Thames regularly froze over in London, though in general summers were similar to the present. Conversely, the period from 100–500 seems to have been unusually warm in Europe. The Romans grew vines in northern England during their occupation.

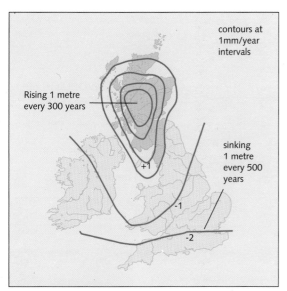

contours at
1mm/year
intervals

Rising 1 metre
every 300 years

sinking
1 metre
every 500
years

+1

-1

-2

Tilting of the British Isles

Britain has experienced a strange set of motions since the end of the last ice age. The north of the island was covered in thick ice. As the weight of ice-sheet withdrew the whole continent of northern Europe gradually rose upwards.This is reflected in raised beaches (effectively the previous sea level), like these on the Isle of Arran,

Scotland (*right*).
In the south of the British Isles sea level is changing, again because of melting ice caps, though this time they are having the opposite effect. The broad deep estuaries of south-west England, as at Salcombe (*above right*) are old river mouths that have been flooded by rising sea levels. Silting is very light, making them ideal natural harbours.

Future World
The Earth continues to change

The Earth's continents are moving, its sea floors are spreading and mountains are being built and eroded. Sediments are being deposited, volcanoes are erupting, and earthquakes are occurring. All of this has been going on for more than 2,500 million years – and will continue for many millions of years to come.

In the geological short term, we can predict continental movements which will continue the motions that are already in place. But after about 50 million years into the future it is unrealistic to give predictions. The Earth's crust is an unstable place and changes in direction of currents in the mantle are only recognized when unexpected events take place on the surface. Small changes in direction of movement can have huge effects. Geologists are engaged in disentangling the evidence of the past. Predicting the distant future is a different task altogether.

One thing we can be sure of is that the heat sources inside the Earth – the radioactive elements – will eventually run down. The Earth's mantle will then stop circulating, will cool and become solid. The plates on the crust will stop moving and fuse into one solid cold plate. Though surface processes, such as erosion and deposition of sediment, will continue for some time, the Earth will become geologically dead, as the Moon has been for most of its existence.

The effects on the surface would be profound. Surface water is constantly lost by the dipersal of hydrogen. The loss is replaced by volcanic eruptions, which bring water from the mantle. When this no longer happens, the surface water would gradually disappear. Cooling of the core would stop the generation of the Earth's magnetic field. The ionosphere would then be dispersed, allowing harmful radiation to reach the surface. Under these conditions it is hard to see how life would survive.

It is natural for human beings to see the world in terms of our own interests, our own history and our impact on the planet. But, though we have undoubtedly affected some of the life-forms and environments of the Earth, we are a small factor in its overall history. There are many metaphors for demonstrating the insignificance of humanity in the

Earth's history; depict the geological timescale as the distance from your nose to the fingertip of your outstretched arm, file off the tip of your fingernail and you have erased all human history.

As we have seen throughout this atlas, the geological activity of the planet is the basis of all surface processes, including the development of life. Even on a geological time scale the death of the planet is a distant prospect – the Earth will remain active for hundreds of millions of years to come.

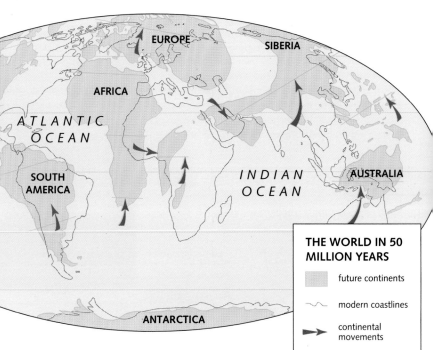

The world in 50 million years
North and South America are turning anticlockwise. How much they remain locked together depends on the relation between the plates around the Caribbean basin.

The biggest split will be along a line running through the Great Rift Valley of East Africa, north-east to the Red Sea, along the Red Sea and the east Mediterranean coasts, and then in a line north-east to Lake Baikal and north to the Arctic coast of Siberia. This will be the centre of the formation of a great new ocean. Although new, this ocean's position will resemble that of the Tethys Ocean which existed from 380–50 million years ago.

The continents to the south of this new ocean will rotate anticlockwise while moving north. The bulk of Africa and Europe will also move northwards, though Europe will also be rotating clockwise. Australia, together with New Zealand and Papua New Guinea will move to the north.

The Earth: present day
Heat sources in the mantle maintain the circulation of convection currents. These recycle plates on the crust and push out steam and other atmospheric gases. Circulation in the outer core maintains the Earth's magnetic field.

THE WORLD IN 50 MILLION YEARS

	future continents
	modern coastlines
	continental movements

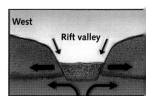

Birth of a new ocean: present-day
The two parts of Africa are being forced apart by convection currents in the upper mantle. As gaps are made, sections of crust drop down to fill them. The classic rift valley shape is the result of this, with transform faults at either side of a flat-floored valley or *graben*.

The future of the Sun and its Solar System

The Sun was formed in the early life of the universe, around 10 billion years ago. At some time the hydrogen that produces its energy will begin to run down. The Sun will expand into a Red Giant temporarily and then collapse into what is known as a White Dwarf. This will continue to shine feebly for a few million years before fizzling out to become a Black Dwarf. By that time, the Earth and the rest of the Solar System will have dispersed as the gravitational pull of the Sun diminishes. The Red Giant phase will come in 5 to 10 billion years. By that time the Earth will have been a dead planet for many millions of years.

The future of the universe itself is a matter of great speculation. It might go on expanding indefinitely, or it might at some point begin to contract, leading to a Big Crunch, and then perhaps another Big Bang.

The Earth: 6 billion years' time

The heat runs out and the mantle stops circulating, then solidifies. The crust forms a thick solid plate. Atmospheric gases and seawater disperse. The planet is geologically dead.

The Earth: 5 billion years' time

As heat sources run down, cirulation in the outer core ceases, eventually solidifying. Lack of a magnetic field disperses the ionosphere, allowing harmful radiation to reach the Earth's surface. Convection in the mantle slows, and fewer plates cover the surface crust.

upper mantle	oceanic crust
continental crust	seawater

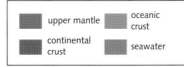

5 miilion years in the future

When the gap between the continents is too wide to be filled by continental crust, the upper mantle is exposed to the surface. It therefore cools and soldifies to form oceanic crust, which is thinner and heavier. This forms the floor of the new ocean, as seawater floods into the depression.

10 million years in the future

A mature ocean has developed with a central spreading ridge, and large flat abyssal plains on either side. Spreading will continue until halted by other plate movements or by changes in the motions of convection currents in the Earth's mantle.

The Great Rift Valley

The fault lines where the plates are moving apart in the Great Rift Valley/Red Sea region are easily visible on a topographic map (*above*). When the continental plates are forced far enough apart for the ocean floor to appear, sea water will flood in and a new ocean will be born.

The Geological Timescales
The Tectonic History and Development of Life

Tectonic History	Development of Life
Pre-Archean 4,600 million years ago Formation of the Earth, Moon and planets. Magnetic field formed at some time in this period.	
PRECAMBRIAN	
4,000 million years ago	
Archean 3,900 million years ago First permanent crust formed. 3,500 million years ago Atmosphere and seawater formed. 3,000 million years ago Greenstone belts – strips of micro-continent.	3,300 million years ago Oldest sedimentary rocks. First stromatolites. Atmosphere with some carbon dioxide. 3,100 million years ago More developed algae and bacteria. 2,900 million years ago Massive stromatolites formed by photosynthesizing blue-green algae.
2,500 million years ago	
Proterozoic 2,500 million years ago Build up of free oxygen in atmosphere. 2,300 million years ago First large-scale glaciation. 2,000 – 1,000 million years ago Rapid growth of continents by accretion of micro-continents. Possible formation of a super-continent. Southern continents combine into Gondwanaland. 700 – 600 million years ago Major glaciation, affecting every continent.	2,200 million years ago Stromatolites common. Atmosphere contains free oxygen. 1,800 million years ago Diversificiation of species of prokaryote algae (cellular forms with no nucleus). 1,400 million years ago Bacteria formed into colonies – first step towards multicellular organisms. Atmosphere rich in oxygen. 1,200 million years ago Development of eukaryote cells. These cells have a nucleus containing DNA, and the capacity for sexual reproduction. 800 million years ago Evidence of sexual reproduction in eukaryote cells. Filamental and tubular algae. Appearance of fungi. 600 million years ago Appearance of diverse species of soft-bodied, multicellular organisms (Ediacaran Fauna).
550 million years ago	

Tectonic History	Development of Life
PALEOZOIC	
550 million years ago	
Cambrian 550 million years ago Laurentia and Baltica positioned in tropics. Gondwanaland stretches from 50°N to the South Pole. Volcanic episodes in the Caledonian region.	550 million years ago Worldwide emergence of marine invertebrate groups with shells and skeletons. Trilobites, brachiopods, archeocyathids, echinoderms, molluscs all common. Stromatolites decline in abundance.
500 million years ago	
Ordovician 500 million years ago Baltica drifts closer to Laurentia – separated by the first Iapetus Ocean. 450 million years ago Taconic Orogeny in north-east Laurentia, caused by collision on offshore island arc.	480 million years ago First definite vertebrates – jawless freshwater fish. Freshwater plants assumed to be present. 450 million years ago Possible first land plants.
440 million years ago	
Silurian 425 million years ago Caledonian Orogeny begins, as Baltica and Avalonia collide with Greenland and Laurentia. Baltica and Laurentia drift near to the African part of Gondwanaland. They are separated by an early version of the Tethys Sea.	440 million years ago Abundance of jawless fish. First fish with jaws – freshwater acanthodians. Giant sea-scorpions (eurypterids) emerge. 420 million years ago First land plants. Vascular plants including lycopsids and psilopsids present, but very rare. First insects and arachnids.
410 million years ago	
Devonian 400 million years ago New phases of the Caledonian disturbances as Gondwanaland rotates clockwise and collides with the eastern margin of Laurentia. The Tethys Sea opens up. 360 million years ago New disturbances along the Gondwanaland/Laurentia boundary, in the final phase of the Caledonian Orogeny. Siberia is the only major block not connected with the Laurentia/Baltica/Gondwanaland landmass.	400 miillion years ago Age of fishes. Jawed and armoured fish become abundant and diversify. Development of modern types of fish with bony skeletons and scales. Spore-bearing plants become more common on land – though still tied to aquatic habitats. 370 million years ago The first amphibians develop from fish and reach the land. Emergence of sea ferns, while true ferns cover some lowland areas in dense forest.
350 million years ago	

Tectonic History	Development of Life
350 million years ago	
Mississippian 350 million years ago Laurentia and Gondwanaland remain associated, though separated by ocean as sea levels rise. Widespread limestone formation.	340 million years ago First true reptiles. Emergence of distinct floras associated with different climatic conditions. Glossopteris flora dominates Gondwanaland.
315 million years ago	
Pennsylvanian 310 million years ago Renewed contact between Gondwanaland and Laurentia causes the start of the Appalachian Orogeny. Gondwanaland has continued to turn clockwise. A major glaciation begins to cover large parts of the southern continents in ice. The Hercynian Orogeny results from the collision of northern Gondwanaland and northern Europe.	300 million years ago Development of huge lycopsid plants in swamp forests. Amphibians and reptiles diversify in humid tropical conditions, as do insects. Abundance of giant flying insects and cockroaches.
290 million years ago	
Permian 270 million years ago Angaraland (Siberia and Kazakhstan) begins to collide with Baltica, creating the Urals. Last part of supercontinent of Pangea is in place. Pangea stretches from 60°N to the South Pole.	270 million years ago As conditions become drier and hotter, reptiles thrive at the expense of amphibians. Development of warm-blooded reptiles (therapsids) the precursors of the mammals. 250 million years ago Mass extinction of marine life. Groups made extinct include trilobites, rugose corals and crinoids. Other marine invertebrates severely affected. Fish are generally unaffected.
250 million years ago	

(vertical label, right margin: CARBONIFEROUS)

MESOZOIC

Tectonic History	Development of Life

250 million years ago

Triassic
250 million years ago
Pangea moves north to straddle the Equator. Many of the continents are now in warm, arid climates. Asian micro-continents begin to move away from Australia and Gondwanaland.

250 million years ago
Ammonites survive the mass extinction at the end of the Paleozoic and thrive in the Mesozoic. Development of thecodont reptiles which become dominant.

220 million years ago
Dinosaurs develop from thecodont reptiles. First mammals emerge from warm-blooded therapsid reptiles. Archeopteryx, the earliest known bird (or feathered dinosaur), develops.

205 million years ago

Jurassic
180 million years ago
Africa and South America begin to split from North America, opening up the Central Atlantic.

150 million years ago
Formation of the Rocky Mountains begin.

210 – 145 million years ago
Dinosaurs become dominant, reaching their largest size. Development and diversification of flying reptiles (pterosaurs) and aquatic reptiles (plesiosaurs). Birds develop and spread widely. Continued diversification of insects.

145 million years ago

Cretaceous
120 million years ago
Africa moves further south, opening a split with Europe. India splits from Africa and Antarctica and begins to move north. Australia splits from Antarctica as Gondwanaland starts to break up.

100 million years ago
South America and Africa begin to split apart – the first time they have been separated since the Precambrian period.

85 million years ago
The central Atlantic stabilizes and links to the still-opening South Atlantic. Changes in Atlantic and Pacific sea floor spreading push Central America and South America together. South America approaches North America, with a narrow ocean basin being squeezed between them. The Andean region becomes a subduction zone.

145 – 65 million years ago
Continuing dominance of land by dinosaurs. Mammals remain small. Reptiles diversify – turtles, snakes, lizards are abundant. Emergence of flowering plants (angiosperms). These dominate the land plant kingdom by the end of the Cretaceous.

65 million years ago
Mass extinction of marine and land life-forms. Principal casualties are the dinosaurs and ammonites.

65 million years ago

TERTIARY

Tectonic History	Development of Life

65 million years ago

Paleocene
Continued uplift in western North America.

65 million years ago
Reptile groups (other than dinosaurs) survive the mass extinction. Mammals and birds also survive and flourish. Emergence of early horse, elephant and bear groups of mammals. Compositae family of plants emerge.

53 million years ago

Eocene
40 million years ago
Uplift of the Rockies and formation of the west coast mountains completed.

50 million years ago
Grasses emerge and diversify rapidly along with Leguminosae and Compositae plants.

40 million years ago
Grazing animals and monkeys emerge. Mammal groups (whales, dolphins) return to the sea. Foraminifera grow and diversify.

36 million years ago

Oligocene
30 million years ago
Japanese islands split from Asia, opening up the Japan Sea.

25 million years ago
Northern North Atlantic opens between Greenland and northern Europe. Africa moves north to close the Tethys Sea and collide with Europe. The Alpine Orogeny continues for 15 – 20 million years.

35 million years ago
The first apes emerge. Large mammals and birds spread over the Earth. Grasses cover large areas of land.

23 million years ago

Miocene
20 million years ago
India begins to collide with Asia in the Himalayan Orogeny.

15 million years ago
Outpourings of basalt lavas in southern Siberia (Baikal Rifts), Central Europe (Rhine Graben), East Africa and Antarctica. Rifts begin in East Africa – first stages in the creation of a new ocean.

10– 11 million years ago
Separation of great apes and hominid apes. Radiation of hominid primates culminates in *Sivapithecus* – an ape showing many characteristics of living apes and humans.

5 million years ago

Pliocene
3 million years ago
Antarctica isolated as South America moves away – the last pieces of Gondwanaland break apart. The Earth's climate cools dramatically.

3–4 million years ago
Emergence of *Australopithecus*. First hominids.

2 million years ago

QUATERNARY

Tectonic History	Development of Life

2 million years ago

Pleistocene
2 million years ago
Nebraskan glaciation (North America) Donau glaciation (Europe).

700,000 – 600,000 years ago
Aftonian interglacial period.

600,000 years ago
Kansan glaciation (North America) Gunz glaciation (Europe).

500,000 years ago
Yarmonth interglacial period.

400,000 years ago
Illinoian glaciation (North America). Mindel glaciation (Europe).

200,000 years ago
Sangamonian interglacial period.

70,000 – 10,000 years ago
Wisconsinan glaciation (North America) Würm glaciation (Europe).

18,000 years ago
Peak of last glaciation.

2–1.75 million years ago
Homo habilis, a possible early form of *Homo erectus*, emerges in East Africa.

1 million years ago
Homo erectus disperses from Africa as far as China and Java.

500,000 years ago
Homo sapiens appears in Africa and migrates to Europe.

250,000 years ago
Modern humans (*homo sapiens sapiens*) emerge in southern Africa.

50,000 years ago
Modern humans reach the Middle East.

35,000 years ago
Modern humans reach Europe as Cro-Magnon man.

30,000 – 20,000 years ago
Modern humans enter North America via the Bering land bridge and move south.

10,000 years ago

Holocene
Continued rifting in East Africa indicates future sea floor spreading zone.

10,000 years ago
Homo sapiens sapiens reaches every continent except Antarctica, and is the only surviving hominid.

| | 550 | 410 | 250 | 65 | 2 |

Glossary

Abyssal Plain
Flat low-lying area of the ocean floors. They lie between the ocean ridges, where crust is created, and ocean trenches where it is destroyed. These are the temporarily stable portions of ocean crust.

Albedo
The proportion of the Sun's radiation that is reflected by any surface. The high albedo of ice- sheets increases cooling in polar regions, once glaciation begins. The reverse is also true – melting of the ice-caps is speeded up by reduced albedo.

Algae
Generally aquatic plants that were among the first celled organisms to evolve. Blue-green algae, or cyanobacteria, use photosynthesis to manufacture food – absorbing carbon dioxide and producing oxygen in the process. They were important in the development of an oxygen-rich atmosphere, allowing further life-forms to emerge.

Alpine Orogeny
An episode of mountain-building volcanism and disturbance which began 25 million years ago, and was caused by the collision of Africa and Europe. This created the Mediterranean Basin and coincided with the Himalayan Orogeny. Africa is still moving towards Europe, causing volcanic activity and earthquakes in Italy and the eastern Mediterranean.

Amino acids
Organic compounds that are the constituents of proteins. An essential stage in the formation of living organisms.

Ammonite
A group of cephalopod molluscs that are the most important fossils of the Mesozoic era. Ammonites are part of the larger ammonoid group, which was almost made extinct at the end of the Paleozoic. One family survived, and the ammonites radiated rapidly. They have coiled shells and are free-floating. Suture lines on their shells are used as markers. Ammonites became extinct at the end of the Mesozoic 65 million years ago.

Angaraland
The ancient continent that comprised present-day Siberia, Kazakhstan and latterly Tarim. Angaraland collided with Laurentia at the end of the Paleozoic era 250 million years ago – the resulting disturbance created the Ural Mountains. The combined continent of Laurentia and Angaraland is known as Laurasia.

Angiosperm
The flowering plants, including all deciduous trees and grasses. Eighty-five per cent of all land plants are angiosperms which emerged in the Cretaceous.

Anoxic
An environment entirely without oxygen. There is still a dispute over whether the Earth's earliest atmosphere was anoxic. Most of the planets and the Moon have atmospheres with no oxygen.

Anticline
The upward portion of a fold, forming an inverted 'U'. Anticlines are important structures as they often act as oil traps. The world's largest oil fields in Saudi Arabia are found in long anticlines.

Appalachian Orogeny
Disturbance caused by the repeated collision of Africa with the south-eastern margin of North America (Laurentia) between 350 and 250 million years ago. The Appalachian Mountains are mainly the result of overthrusting, as huge quantities of sedimentary rock were pushed westward by the collision. Florida became joined to America at the end of the Appalachian Orogeny.

Aragonite
Form of calcium carbonate often found in the shells of marine animals, and in stalactites.

Archean
First period of the geological time-scale and part of the Precambrian era, lasting from the oldest known rocks (c. 3,900 million years ago) to 2,500 million years ago. The boundary with the younger Proterozoic period is difficult to define, except by radioactive dating, because of the great age of the rocks and the lack of fossils.

Archeocyathid
An important group of shelled marine invertebrates which developed rapidly in the Cambrian, and became extinct at the end of the period. Fossils have been found in Cambrian rocks on every continent.

Archeopteryx
One of the most important fossil finds, Archeopteryx was a feathered dinosaur, half-way between a dinosaur and a bird, and therefore proof of the development of one from the other. Fossils, showing perfect traces of feathers, have been found in Triassic rocks in southern Germany.

Asteroids
Small objects, mainly the left-over pieces of planetary formation, which circulate in the solar system – but especially in the asteroid belt between Mars and Jupiter.

Avalonia
A small continent comprising the southern British Isles and perhaps part of Nova Scotia and Newfoundland in the early Paleozoic era. Collision with Baltica, Laurentia and Greenland caused mountain-building on its northern and western margins – as seen in the mountains of Scotland, Wales, Northern Ireland and the maritime provinces of Canada.

Baltica
A small continent comprising present-day Scandinavia and the region around the Baltic Sea. It collided with the British Isles and with Laurentia in the early Paleozoic. Siberia joined on to Baltica in the late Paleozoic, and this then became a stable region.

Basalt
Fine-grained igneous rock with a particular

composition. Oceanic crust has a similar chemical composition. Basalt flows on the continents are often associated with splitting apart of the continents (rifting), and the opening of new oceans.

Big Bang
The theoretical beginning of the universe. A gigantic explosion produced prodigious quantities of energy, some of which converted into matter. If Big Bang was the beginning of everything, then there was nothing to cause it. Some interpretations of cosmology and relativity theory suggest that time itself did not exist before Big Bang.

Big Crunch
A theory which suggests that the present expansion of the universe will implode into a Big Crunch, before another Big Bang sets the cycle off again.

Bivalve
Aquatic molluscs with two identical shells, e.g. present-day oysters and mussels. Used locally as zone fossils, they originated in the Lower Cambrian.

Black Hole
The death of a star can result in an implosion, where the gravitational force is enough to prevent any light from passing out of the surrounding region. This then becomes a black hole.

Blastoid
A marine invertebrate of the echinoderm group, with a characteristic neo-spherical shape.

Brachiopod
One of the most important fossil groups in the geological record. They had two shells of unequal size and were attached to the sea-floor by a stalk. Variations in shell form allow them to be used as zone fossils in the Ordovician, Silurian, Carboniferous and Cretaceous periods. They evolved in the Lower Cambrian, and are still present.

Calcareous
Used to describe any rock containing calcium carbonate.

Caldera
A volcanic crater causd by the collapse or emptying of the underlying magma chamber during and after eruption, which can become filled with water to form a crater lake.

Caledonian Orogeny
The first great geological disturbance of the Paleozoic era. Collisions between different parts of northern Europe and Greenland created the mountains of northern Britain and Scandinavia.

Cambrian
The first geological period of the Paleozoic era, lasting from about 550 to 500 million years ago. The beginning is defined by the appearance of large numbers of shelly fossils. Cambria is the ancient name for Wales, where much early work was done on rocks of the period.

Capping Rock
An impervious layer of rock (usually salt, shale or evaporite), lying on top of oil and gas deposits, preventing them from escaping to the surface.

Carbonate Rock
Any rock composed principally of carbonate minerals – usually limestone or chalk, (calcium carbonate) or dolomite (calcium magnesium carbonate). Carbonates are chemically precipitated from sea water.

Carbon Cycle
Mechanism by which the carbon content of seawater, atmosphere and rocks is kept in balance. Carbon dioxide in rainwater forms carbonates as it washes limestones off the land into the sea. As seawater becomes saturated, carbonates are deposited on the floor to form more limestones (later to be uplifted onto land), while some excess carbon dioxide is evaporated back into the atmosphere.

Carboniferous
The geological period lasting from about 360 to 290 million years ago. It is divided into Upper and Lower Carboniferous (or Mississippian and Pennsylvanian) because of a great change in rock types in the middle of the period. The period is named after the coal-rich deposits of the Upper Carboniferous.

Cenozoic
The era covering the last 65 million years of the Earth's geological history, including the Tertiary and Quaternary periods.

Cephalopod
A class of marine molluscs which includes ammonites and belemnites. The shell is often multi-chambered and used for buoyancy. Living members include squid, octopus and cuttlefish.

Chalk
A pure, fine-grained, soft form of limestone, composed almost entirely of the bodies of microscopic marine organisms. Chalk rock formations indicate an ancient environment of warm, tranquil, shallow seas.

Chemical Sedimentation
When seawater or lake water becomes saturated in certain minerals, it will be deposited on the sea floor – either as an ooze, or via the shells of marine animals. These are sometimes formed into sedimentary rocks – usually carbonates.

Chordate
An animal with a backbone, a member of the chordata group.

Clastic
Sedimentary rock (sandstones, shales, conglomerates) formed by physical deposition – generally by solid particles falling through water. Constituent particles will have been broken off other rocks by erosion.

Coal
Sedimentary rock made of decomposed plant remains. The carbon content of the plants is retained by decomposition in the absence of oxygen. Heat and pressure during and after formation increases the fuel value of the coal.

Coccolith
Microscopic plates of calcium carbonate formed by the death of planktonic organisms. Coccoliths fall to the sea-floor in billions to eventually form chalk. They are

formed only in warm, tropical seas.

Comet

A small body usually made principally of ice and dust with an eccentric orbit that sometimes brings it into the inner solar system. Comets may be once-only visitors to the inner solar system. The characteristic tail is made of dust particles.

Conglomerate

Sedimentary rock made of large boulders and fragments, usually embedded in a fine matrix. Conglomerates are formed in extremely turbulent conditions – rapid streams, landslides, avalanches.

Continental Drift

The movements of the continents across the surface of the Earth. The expression was coined before any mechanism for the moving of continents was known.

Convection Cell

Convection works because hot materials expand, become less dense, and float up through colder more dense matter. In the Earth's mantle material is heated at depth and rises. Convection cells form as the rising material cools at the surface and falls back down again. The cells are reflected in regular motions across the surface – and it is these that are thought to push the plates of the crust around (see also Plate Tectonics).

Cordilleran Orogeny

A series of disturbances resulting in the uplift of the Rocky Mountains and the formation of the Sierra Nevada and Coastal Ranges.

Core

The inner part of the Earth, beginning at a depth of about 2,900 km (1,800 miles), as defined by seismic studies. The outer core is probably liquid iron with about 10 per cent nickel. Currents in the outer core drive the Earth's magnetic field. The inner core, beginning at a depth of 5,100 km (3,169 miles) is solid and may be crystallizing out of the liquid outer core.

Cosmology

The study of the cosmos, or universe.

Craton

The stable ancient centre of a continent. This term is generally used to describe the Archean or Precambrian parts of the continents formed before the orogenic episodes of the Paleozoic era.

Cretaceous

The last period of the Mesozoic era, lasting from 140 million years ago to 65 million years ago. The Cretaceous period is named after the thick chalk deposits on the northern continents. The end of the period saw the extinction of the dinosaurs and ammonites.

Crinoid

Echinoderms which are attached to the sea floor by a long stalk, or are free-floating. They were an important fossil group in the Paleozoic.

Crust

The solid outer layer of the Earth. The base of the crust is defined by a change in the behaviour of seismic waves as they pass into the more plastic upper mantle. Continental crust is on average 35 km (21 miles) thick and is comparatively light. Oceanic crust is on average 5 km (3 miles) thick and more dense. The plates of the crust effectively float on the mantle. The crust is less than 0.1% of the Earth by volume.

Curie Point

The temperature above which materials lose their magnetism – for rocks this is about 500°C. The Curie point is important since igneous rocks are magnetized on cooling. Reheating of rocks in orogenic disturbances can remagnetize them.

Cyclothem

A cycle of deposition which shows changing coastal conditions. Limestone, shale, sandstone and coal measures occur in repeated bands (though not always with all present), indicating the change from shallow marine, to estuary, to swamps, as a result of changing sea levels.

Devonian

The geological period from 410 to 350 million years ago. Known as the Age of Fishes, this period also saw the extensive establishment of land plants. It is named after extensive formations in the county of Devon in south-west England.

Dinosaur

Members of two orders of reptiles – the saurischians and ornithischians. They originated in the Triassic and dominated the land in the Jurassic and Cretaceous periods. Their extinction at the end of the Cretaceous, 65 million years ago, is one of the great puzzles of Earth history.

DNA

Dioxyribonucleic acid is the chemical molecule that carries genetic information from one generation to the next. Errors in DNA copying lead to mutations, essential to evolution. The complex double-helix structure of DNA was discovered in the 1950s.

Dolomite

Mineral with the chemical composition $CaMg(CO_3)_2$. It may be formed by the circulation of magnesium-rich sea water through limestone.

Earthquake

A disturbance caused by the relative motion of crustal plates. Earthquakes occur at or around plate boundaries.

Ediacaran

Fossil types named after those found in the Ediacaran hills of South Australia. They are complex soft-bodied animals which were found in rocks c.650 million years old.

Epicentre

The point on the Earth's surface vertically above the centre, or focus, of an earthquake.

Eucaryote

A single-celled animal with a nucleus. The development of a nucleus with DNA allowed a sexual reproduction to begin. Eucaryotes emerged 800 million years ago.

Eurypterids

Sea-scorpions from the Paleozoic era. Eurypterids grew to lengths of 3 metres (10 ft). They became extinct at the end of the Permian period.

Evaporite

Mineral formed by evaporation of lakes or tidal flats under heat from the sun. Evaporites are indicators of tropical or arid conditions.

Evolution

Accidental mutations arise in species. When these are favourable to life, they will persist as more mutated individuals will live to reproduce and pass on the mutations. This natural selection is the basis of evolution.

Exoskeleton

The external skeleton which first developed in the Cambrian period in certain groups of marine invertebrates, providing muscle anchorages and protection from predators.

Fault

A fracture in brittle rock caused by movement in opposing directions. The surface along which movement occurs is known as the fault plane. Large-scale movements can provide faults running for thousands of kilometres. Thrust faults occur when the rocks on the lower side of the fault plane are pushed up and over those on the upper side. In a fault zone numerous faults run in almost parallel directions.

Fauna

Members of the animal kingdom. The term is usually used to describe a group of animals existing in one place and/or time.

Faunal/Floral Exchange

When continents move together, previously isolated groups of plants and animals can migrate between the two – though not all of them will.

Faunal Realm/Province

A region where a particular set of animal groups or species exists or existed. The boundaries are defined by changes in animal types.

Flora

Members of the plant kingdom. The term is used to describe the group of plants existing in a defined place and/or time.

Floral Realm/Province

Plant equivalent of Faunal Realm/Province.

Fold

A bend or buckle in existing rock, caused by external pressures. The upward arch is known as an anticline, the downward curve is a syncline.

Foraminifera

Minute aquatic organisms with a single or multi-chambered shell, which are important age indicators in various geological periods.

Fossil

The preserved remains of a plant or animal. Actual parts of the organism may be preserved – normally the shell or exoskeleton. More usually hard or soft parts are replaced during lithification by minerals, which are sufficiently different to the surrounding rock for the presence of the organism to be preserved.

Galena

Lead sulphide – the most important lead ore.

Geology

The study of the Earth, including life-forms, throughout its history.

Geosyncline

Trenches created within active subduction zones are often filled with sediment eroded from newly formed mountain chains. Vast quantities of sediment can build up as the crust is pushed down under its weight. When lifted, geosynclines are among the most important geological formations.

Geyser

A vertical jet of steam ejected by groundwater coming into contact with hot rocks under the surface. Geysers are common in certain active tectonic zones, for example Iceland, New Zealand.

Glaciation

Cooling which is sufficient to form ice-sheets and glaciers. Episodes of glaciation occur throughout the Earth's history, but the presence of polar ice-sheets (as at present) is relatively uncommon.

Glossopteris

Fossil ferns from the Late Paleozoic, found on all the southern continents – then grouped together in the super-continent of Gondwanaland.

Gondwanaland

The southern continent that comprised South America, Africa, Antarctica, Australia, India and parts of southern Asia. Gondwanaland existed from the Precambrian until the Jurassic period – for the vast majority of the Earth's geological history. The existence of Gondwanaland was a source of great dispute among geologists until the 1960s.

Granite

Coarse-grained igneous rock containing grains of quartz, mica and feldspar.

Graptolite

An important fossil group in the Paleozoic era, used as zone fossils in the Ordovician and Silurian periods. Graptolites secrete a branching exoskeleton, which leaves impressions in shales.

Greenstone Belt

Archean formations containing some of the world's oldest rocks. These elongated belts are thought to reflect the shapes of the earliest strips of continent.

Gymnosperm

A seed-bearing plant – e.g. conifer. Gymnosperms have a more efficient reproductive system than their predecessors, the spore-bearing plants, but declined with the later emergence of angiosperms, the flowering plants.

Hard Parts

Animals first started to manufacture hard parts, shells and skeletons, at the start of the Cambrian period 550 million years ago. This lead to abundant fossilization. Calcium carbonate is the principal constituent of animal hard parts.

Hercynian Orogeny

A disturbance caused by the collision of Africa and northern Europe in the Upper Paleozoic. Large-scale igneous activity resulted. The Ardennes, Harz Mountains and Bohemian Massif date from this episode.

Holocene
The most recent period of geological history, lasting from 10,000 years ago to the present.

Hominid
The direct ancestors of modern humans which emerged 4 to 5 million years ago in Africa.

Hydrothermal
Literally 'hot water'. Hydrothermal fluids carry minerals in solution, and deposit them on cooling. Hydrothermal veins are often rich sources of economic minerals.

Iapetus Ocean
The forerunner of the Atlantic, the Iapetus Ocean lay between Laurentia and northern Europe. The Iapetus closed during the Caledonian and Appalachian Orogenies.

Ice Age
A term loosely used to describe an intense period of glaciation. The peak of the last ice age was about 18,000 years ago. Though we are now between ice ages, this is still an unusually cold period in the Earth's history.

Ichthyosaur
A marine reptile which existed at the same time as the dinosaurs, its close relatives.

Igneous
Rocks formed from molten magma originating deep in the crust or mantle. If magma erupts onto the surface the resulting rocks are volcanic. If magma solidifies beneath the surface it forms plutonic or intrusive rocks.

Index Fossil
See *Zone Fossil*

Ionosphere
A layer of the atmosphere, which offers protection from the Sun's radiation. It is held in place by the Earth's magnetic field.

Island Arc
The characteristic shape of an island chain formed by the subduction of one oceanic plate beneath another. Present-day examples include the Aleutian islands, Lesser Antilles, Sumatra and Java.

Isostasy
The way in which the solid continents float on the plastic mantle.

Jurassic
Middle period of the Mesozoic era, lasting from 210 to 140 million years ago.

Land Bridge
A temporary link between two land masses caused by lowering of sea levels. Land bridges allowed the migration of land-based animals between otherwise separated continents.

Laurasia
A continent comprising Laurentia (North America and Northern Europe) and Angaraland (Siberia and Kazakhstan), which became joined 250 million years ago.

Laurentia
The name for the ancient North American continent, and for the expanded continent after it became joined to Baltica and northern Europe.

Lava
Molten magma becomes lava when it reaches the surface. Different chemical compositions produce a range of flow patterns – some lavas are erupted explosively, while some flow smoothly from tissues in the crust.

Limestone
A sedimentary rock, made almost entirely of calcium carbonate. Limestones are formed mainly from the calcite shells of marine animals.

Lithification
The formation of rock from sediment. The process involves compression and hardening; water is squeezed out and chemical changes are induced, resulting in the formation of rock strata.

Lithosphere
The crust, together with the rigid upper layer of the mantle.

Lungfish
A type of fish that developed lungs alongside gills in the Devonian period. Lungfish are seen as the precursors to amphibians, and therefore to the emergence of vertebrates onto land.

Lycopsid
A group of plants including the species Lepidodendron and Sigillaria that dominated the swamp forests of the northern continents during the Carboniferous period.

Magma
Molten material generated deep in the crust or upper mantle. Magma makes its way up through the crust, where it may remain molten in magma chambers before erupting onto the surface. Alternatively, magma may cool and solidify before it reaches the surface forming batholiths; or erupt directly onto the surface.

Magnetic Inversion
The reversal of the Earth's magnetic field, probably caused by changing currents in the outer core. Reversals are recorded in rocks, giving characteristic magnetic patterns on ocean floor rocks.

Mantle
The interior part of the Earth extending from the crust to the outer core, 2,900 km (1,800 miles) below the surface. The upper mantle is fairly rigid to a depth of about 400 km (250 miles) – though with considerable variations. The lower mantle is molten. Convection currents in the mantle cause continental movement.

Mantle Plume
A hot rising current in the mantle producing volcanic activity on the surface. The Hawaiian Islands sit on top of, and owe their existence to, a mantle plume.

Mare
The 'seas' of the Moon – large flat areas once thought to be water. They are in fact lava flows.

Marsupial
Mammals that give birth to underdeveloped young which are then nurtured in pouches in the mother's body. They spread through the southern continents, and became

isolated in Australia and South America as Gondwanaland split apart.

Mesozoic
The 'Middle life' era lasting from 250 to 65 million years ago, containing the Triassic, Jurassic and Cretaceous periods. This was the Age of the Dinosaurs – the upper boundary is marked by their extinction.

Meteorite
Large extra terrestrial masses that reach the Earth's surface. Meteorites have similar composition to the terrestrial planets; they are probably left over pieces of planetary material that never accreted on to the planets. Their composition is studied for clues to planetary formation.

Milankovitch Cycles
Three cycles which cause changes in the Earth's temperature: variations in the Earth's orbit, the angle of the rotation axis, and the amount of 'wobble' around the axis. The theory accounts for calculated changes in temperature over the past 400,000 years. Other causes, including continental movements, account for longer term variations.

Mineral
Naturally occuring compound with a definite form and crystal structure. Minerals are the constituents of rock.

Mississippian
Alternative name for the Lower Carboniferous period, lasting from 350 to 315 million years ago.

Mollusc
A group of invertebrates including bivalves, gastropods and cephalopods – some of the most important marine fossil groups.

Monotreme
Egg-laying mammals (the duck-billed platypus and echidna) found only in Australia.

Natural Selection
The central mechanism of evolutionary theory. Individuals with favourable mutations survive in greater numbers to reproduce, and pass on their mutations to the next generation.

Neanderthal
The variety of *Homo sapiens* that colonized Europe 300,000 years ago. They were later replaced by modern humans – *Homo sapiens sapiens*.

Nebula
A cloud of gas. The sun and planets are thought to have condensed from a nebula about 5,000 million years ago.

Ocean Ridge
The axis of sea floor spreading where ocean crust is created and pushed sideways. Ocean ridges are active earthquake and volcanic zones.

Ocean Trench
Where an ocean plate is subducted beneath a continent, a trench is formed along the margin between the two. These trenches later become filled with sediment eroded from the mountain chains formed on the edge of the continent. Present day trenches include the Peru-Chile trench alongside the Andes, and the Marianas trench alongside the Philippines.

Oil
Crude oil is formed from the decomposed bodies of minute marine organisms. Being light, oil migrates upwards until it meets a non-porous capping rock.

Oil Trap
Rock formations in which impervious capping rock overlies the porous reservoir rock where the oil is stored. Traps are often formed in anticlines, or in fault systems.

Old Red Sandstone
The rock formation of the Devonian period found across northern continents. It was created by erosion of vast quantities of material from the newly-formed Caledonian Mountains.

Ooliths
Tiny grains of calcium carbonates formed around fragments of shell or grains of sand. Oolitic limestone is made up of ooliths cemented together, and resembles fish eggs in appearance.

Ordovician
The period of the geological timescale lasting from 500 to 440 million years ago.

Organic
Having life, or originating from a life-form. The term is also used to define hydrocarbon chemistry. Confusion arises over organic carbons in the oldest rocks, which may not be evidence of life.

Ornithischian
One of the two great orders of dinosaurs. Ornithischians had bird-like hips, and are almost definitely the precursors of birds.

Orogeny
An episode of great geological disturbance, usually caused by the collision of two or more continents. Orogenies created the great mountain chains of the world, e.g. the Appalachians, Caledonian, Ural Mountains. The Alpine-Himalayan Orogeny is still active.

Outcrop
A rock formation that reaches the surface, and is therefore visible.

Outgassing
The theory that atmospheric gases and seawater were produced in the mantle and erupted onto the surface through volcanic activity.

Overthrusting
The pushing of rock strata up and over younger or similar aged rock. Overthrusting can happen on a continental scale – for example in the building of the Appalachians, the Rockies, and parts of the Himalayas.

Paleogeography
Literally 'ancient geography' – the plotting of the positions of the continents, mountain chains, and oceans at different times through the Earth's history.

Paleomagnetism
The direction of magnetization preserved in ancient rocks is a good indicator of the latitude in which they were formed. Paleomagnetic studies have also revealed the

periodic switching of polarity of the Earth's magnetic field. Paleomagnetism is one of the most valuable tools in the reconstruction of past geography.

Paleontology
The branch of geology specializing in the study of ancient life forms.

Paleozoic
The geological era stretching from the start of the Cambrian period 550 million years ago, to the end of the Permian period 250 million years ago. Paleozoic literally means 'ancient life'.

Pangea
The supercontinent formed at the end of the Paleozoic era 250 million years ago. Pangea consisted of all the major continental blocks, but began to break up almost as soon as it had formed.

Permian
The last period of the Paleozoic era, lasting from 290 to 250 million years ago. The end of the Permian saw the extinction of several families of marine animals, and the formation of Pangea.

Photochemical Dissociation
The theory that the free oxygen in the Earth's atmosphere arose through the dissociation of water vapour, in the presence of sunlight, into its constituent elements; hydrogen and oxygen.

Photosynthesis
The chemical process used by plants to manufacture food using sunlight. Photosynthesizing plants absorb carbon dioxide and produce oxygen. The great expansion in the amount of free oxygen in the Earth's early atmosphere probably came about through photosynthesis by cyanobacteria (blue-green algae).

Physical Sedimentation
The deposition of sediment by solid particles falling through water onto a sea bed, river bed, beach or delta, or after transport by glacier, ice-sheet or wind. Sandstones, shales and conglomerates are all formed by physical sedimentation. *See Chemical Sedimentation*.

Placental
Mammals that give birth to fully developed young. Placentals have superseded marsupials in most parts of the world. The distribution of the two groups is an indication of continental movements over the last 70 million years.

Planetessimal
Small pieces of material that condensed out of the solar nebula. These are then thought to have accreted to form the planets.

Plate Boundary
The boundaries between plates of the crust are the geologically active regions of the Earth. At constructive plate boundaries, new crust is created and the plates are pushed away from each other. At destructive plate boundaries plates are moving towards each other; one is subducted beneath the other, and is subsumed back into the mantle.

Plate Tectonics
The overall theory explaining the mechanisms by which the geological processes of the Earth are driven. Plate tectonics explains how the Earth's crust is divided into rigid plates, how these plates are moved around on the surface by currents in the underlying mantle, and how new plates are continually formed and old ones destroyed.

Pleistocene
The period lasting from 2 million to 10,000 years ago. Notable for the cooling of the Earth, leading to a series of ice ages.

Plesiosaur
Marine reptiles from the Mesozoic era, which were close relatives of the dinosaurs.

Precambrian
The era covering the Earth's geological history from the oldest rocks (about 4,000 million years ago) to the start of the Cambrian period, 550 million years ago. With very little fossil evidence, the Precambrian was beyond the reach of early geologists, who relied on fossils for the comparative dating of rocks.

Protein
Complex chemicals which are the essential constituents of living cells. They are of interest to paleontologists because the formation of proteins was an essential stage in the origin of life.

Proterozoic
The period lasting from 2,500 to 550 million years ago. It was part of the Precambrian era, in which life forms began to be more abundant.

Province
A region bounded by a marked change in fauna, flora or rock type. *See Faunal Province*.

Psilopsid
The earliest vascular plants which colonized the land in the Devonian and Silurian periods.

Pterosaur
A reptile which developed the ability to fly. Pterosaurs were close relatives of the dinosaurs .

Pyroclastic
The material ejected into the air in a volcanic eruption. Pyroclastic debris can travel over great distances before reaching the ground.

Quaternary
The geological era covering the last 2 million years of Earth history.

Radioactive Dating
The measurement of the ratio between different isotopes of a radioactive element can give an accurate estimate of the age of the formation of the host rock, since rates of decay are known for most elements. Radioactive dating has helped geologists to be more precise about dates of rocks and to find out more about rocks where fossil evidence is absent.

Reef
A large structure made of organic material. Ancient reefs are often preserved as rock in the geological reord. Reefs are made of the skeletons of corals, crinoids or other marine organisms. Reef building usually occurs only within a zone around the equator, so ancient reefs are a guide to paleogeography.

Rift Valley

A valley with steep sides and a flat bottom, created by the central piece of land falling below the two sides. In plate tectonics this is caused by the rifting apart of two plates. This is presently occurring in East Africa along the line of the Great Rift Valley.

Richter Scale

The measurement of the intensity of earthquakes, using an assessment of the amplitude of the seismic waves. The Richter scale is logarithmic, so that an earthquake measuring 7 is 10 times more intense than an earthquake measuring 6. There is no upper limit to the scale, but in practice no earthquake registering larger than 8.9 has ever been recorded.

Rotational Poles

The ends of the axis on which the Earth rotates. The rotational poles vary slightly from the magnetic pole. The two are associated because the Earth's rotation influences the currents in the outer core which drive the magnetic field.

Sabkha

A tidal salt flat, particularly on the Persian Gulf. Sabkhas form layers of salt in rock formations and act as oil traps, particularly on the Arabian peninsula.

Salt Dome

A structure caused by the migration of salt, which is light, up through denser overlying strata. Salt domes are important because salt is impervious and often acts as an oil trap.

Sandstone

A common sedimentary rock formed on beaches, deltas, coastal trenches, deserts and a variety of other environments.

Saurischian

One of the two orders of dinosaur. Saurischians have a more reptile-like hip bone than the ornithiscians.

Sea Floor Spreading

The creation of new oceanic crust at a central ridge, which is then pushed outwards on either side. Sea floor spreading is the main mechanism by which tectonic plates are moved around the Earth's surface.

Sea Mount

A submarine mountain. Many sea mounts were previously volcanic islands, but have been eroded to lie just below the surface, for example the Emperor sea mounts north-west of Hawaii.

Sedimentary Rock

Rock, including limestones, sandstones and shales, formed from the deposition of eroded fragments of other rocks (physical sedimentation) or from the precipitation of minerals out of sea water or lakes (chemical sedimentation).

Seat Earth

The layer of soil in which coal swamp plants grew. The seat earth is sometimes preserved as a thin layer of rock underlying a coal seam.

Seismic

A natural or artificially induced earthquake or vibration. Seismic waves are used to study the interior of the Earth.

Shale

A fine-grained sedimentary rock, formed in tranquil conditions – swamps, lakes, calm areas of deltas, mudflats and rivers.

Shooting Star

A small piece of interplanetary material, for example a small meteor, burning up in the Earth's atmosphere.

Silurian

A period of the early Paleozoic era, lasting from 440 to 410 million years ago. It was named after the Silures – an ancient tribe in Wales.

Strata

Layers of sedimentary rock which may or may not be lying horizontally, and are subject to folding and faulting.

Striations

Scratch marks on rocks caused by ice-sheets or glaciers. As the ice passes over the underlying rock, the material embedded in the bottom scrapes lines across it. Striations are parallel, and are definite indicators of previous glaciation and of the direction of ice flow.

Stromatolite

A massive structure comprising blue-green algae interlayered with lime 'mats'. Stromatolites form in intertidal waters. They were common in the Precambrian era, when tides were greater. Fossil stromatolites are among the earliest life forms found on Earth.

Subduction Zone

A region of the crust where one plate is being forced beneath another – usually an oceanic plate beneath a continental plate. Subduction zones are the geologically most active parts of the crust, with mountain-building, volcanism, earthquakes and geosynclines all present.

Super-Continent

A landmass comprising two or more major continental blocks. The term is most often used to describe Gondwanaland and Pangea.

Supernova

A violent explosion resulting from the gravitational collapse of a massive star, when the outer layers are blown away. Some elements in our solar system were probably created in a nearby supernova.

Suspect Terrane

A piece of continental crust that has been transported across a large distance before becoming accreted onto a continental margin. The western margin of North America has many suspect terranes attached to it.

Taconian Orogeny

The equivalent of the early Caledonian Orogeny in North America. It takes its name from the Taconic Range in the north-east United States.

Tectonic

Relating to large scale forces and movements in the Earth. See Plate Tectonics.

Terminal Moraine

The deposit formed at the end of a glacier or ice-sheet, as the ice melts and deposits its load.

Terra

The 'mountains' of the Moon. The term literally means 'land', and was used to differentiate these regions from the flat 'mare', which were thought to be seas.

Terrestrial Planet

The inner planets of the solar system with a similar composition to the Earth; Mercury, Venus, Earth and Mars.

Tethys Ocean

The ancient sea which separated Gondwanaland from Laurentia for most of the Mesozoic era. The Tethys shrank, and then disappeared entirely in the Alpine Orogeny. Its marine fauna is found in fossil beds across Europe.

Tetrapod

A four-legged animal. The first tetrapods were a transitional form between fish and amphibians, and therefore an important stage in the movement of life onto land.

Thecodont

The most important reptile group of the Triassic period. The thecodonts gave rise to the dinosaurs some time in the Late Triassic.

Therapsid

The group of reptiles which developed mammal-like characteristics in the Triassic period. They became extinct in the Mesozoic. It is generally thought that they are the ancestors of the mammals.

Tillite

A type of deposit formed by an ice-sheet. Tillites are mixtures of mud, sand and boulders. Discoveries of ancient tillites in tropical regions led early geologists to consider the possibility of continental movement.

Traps

Large flows of fine-grained igneous rock, for example basalt.

Triassic

The first period of the Mesozoic era, lasting from 250 to 210 million years ago. The first mammals emerged in the Triassic as did the dinosaurs.

Trilobite

The most important fossil group of the Early Paleozoic era. Trilobites have external skeletons, divided longitudinally into three sections or lobes. Variations in trilobite shapes and features are indicators of the age of the host sediments. The trilobites became extinct at the end of the Paleozoic era 250 million years ago.

Tsunami

A tidal wave. Tsunamis are generally caused by earthquakes measuring more than 6.5 on the Richter scale and with a depth of less than 5 km (3 miles). Not all such earthquakes cause tsunamis, and the precise mechanism by which they are created is unclear.

Vascular Plant

A plant with a system for circulating fluids internally. The first land plants had to have ways of distributing nutrients, the task that is done by seawater in marine plants. Vascular plants are therefore an important first step in the development of life on land.

Volcanic

Molten material that is erupted or extruded onto the Earth's surface from deep in the crust or mantle. While some volcanic material is explosively erupted, much is gradually pushed out through fissures in the crust.

Zone Fossil

A fossil group that is used to pinpoint the relative ages of rock strata, especially when these are geographically separated. Zone fossils are usually free-floating marine creatures, with characterisitc variations with age, that are rapidly spread. Also known as index or marker fossils.

Index

Items set in **bold** are pictures, illustrations or maps.

W

X

Y

Z

Select Bibliography

Ager, Derek, *The Nature of the Stratigraphic Record,* 2nd edition (John Wiley, New York, 1980)

Ambraseys, N.N. and Melville, C.P., *History of Persian Earthquakes* (Cambridge University Press, Cambridge, 1982)

Bradberry, J., *Introducing Earth Science* (Basil Blackwell, Oxford, 1985)

British Paleozoic Fossils, British Mesozoic Fossils, British Cenozoic Fossils (British Museum, Natural History, London, 1971–1993)

Darwin, Charles, *On the Origin of Species by Natural Selection* (John Murray, London, 1859)

Dott, R.H. and Batten, R.L., *Evolution of the Earth,* 4th edition (McGraw-Hill, New York, 1988)

Gould, Stephen Jay, *Time's Arrow, Time's Cycle: Myth and Metaphor in the Discovery of Geological Time* (Harvard University Press, Cambridge Massachussets, 1987)

Hallam, A., *A Revolution in the Earth Sciences* (Oxford University Press, Oxford, 1973)

Hamblin, W.K., and Howard, J.D., *Exercises in Physical Geology,* 7th edition (Macmillan, New York 1989)

Hutton, James, *Theory of the Earth* (Cadell & Davies, London, 1795)

Lambert, David, *Dinosaur Data Book* (Facts on File, New York, 1988)

Levin, H.L., *The Earth Through Time* (Saunders, Philadelphia, 1978)

Lyell, Charles, *Principles of Geology* (John Murray, London, 1830–1875)

Porter, Roy, *The Making of Geology: Earth Science in Britain 1660–1850* (Cambridge University Press, Cambridge, 1977)

Press, F. and Siever, R., *Earth,* 4th edition (W.H. Freeman, New York, 1986)

Rupke, N.A., *The Great Chain of History* (Clarendon Press, Oxford, 1983)

Seyfert, C.K. and Sirkin, L.A., *Earth History and Plate Tectonics* (Harper & Row, New York, 1979)

Stanley, S.M., *Earth and Life Through Time* (W.H. Freeman, San Francisco, 1986)

Tarling, D.H., *Paleomagnetism,* 2nd edition (Chapman and Hall, London, 1995)

Tarling, D.H., *Plate Tectonics and Biological Evolution* (Carolina Biological Supply Company, Burlington, 1992)

Wegener, Alfred, *The Origin of Continents and Oceans,* English translation of 4th edition (Dover, New York, 1966)

Acknowledgements

Front cover (clockwise around the title):

Optical photograph of M31, the Andromeda Galaxy: Tony Hallas/Science Photo Library

Australia, Victoria Coast: Images Colour Library Limited

Sprays of hot molten lava from Kilauea Volcano on the Island of Hawaii: Rick Golt/Science Photo Library

Moods of Nature – California, Redwood National Forest: Images Colour Library Limited

West Indies – Virgin Islands, Trunck Bay, St John: Images Colour Library Limited

USA – Arizona, The Grand Canyon: Images Colour Library Limited

Waterfalls: Images Colour Library Limited

Lightning over Landscape: Images Colour Library Limited

American Museum of Natural History, Courtesy Department Library services, Neg. No. 333859 Photo. Rota,: **63br**; Neg. No. 314057 Photo. Rice È Dutchen:**65tl**

Bruce Coleman Limited: Eric Crichon:**112bl**; Jan Taylor:**112br**; Jane Burton: **143bc**

Images Colour Library Limited: **22b, 35tl, 47, 105t, 105b, 111, 135cl, 144t**

Landform Slides: **37cr, 37br, 40, 41bl, 41c, 41tr, 42t, 45tr, 45br, 46, 61, 92bl, 93, 109r, 141cl, 154t, 169b**

Norton W.W./Marianne Collins: **drawing p51**

Professor N.N. Ambraseys/Imperial College of Science & Technology, London: **165tr**

Russian and Republics Photo Library: **87**

Science Photo Library: Alfred Pasieka: **101c, 101bc**; Arnold Fisher: **143bl**; C.Whiddington: **32**; Claude Nuridsany & Marie Perennou: **71tl**; David P. Anderson, SMU/NASA: **23cr**; David Weintraub: **160bl**; Doug Allan: **166t**; Dr David Miller: **166b**; Dr Peter M. Borman, Poroperm-Geochem Ltd: **71tl**; Earth Satellite Corporation: **41br**; Geoff Tompkinson: **135bcl**; Geospace: **22cr**; Ian Steele & Ian Hutcheon: **21tr, 21cr**; J.C. Revy: **78l**; John Farmer: **169t**; John Mead: **79**; John Reader: **157bl**; John Sanford: **20, 24cl**; Martin Land:**78br, 106bl**; Michael Barnett: **78bl**; NASA: **22cl, 22-23t, 22-23b, 23tl, 23br, 25b, 45l, 130, 146tl**; NOAO: **21tl**; Peter Menzel: **101cr, 163tr**; Prof. Stewart: **33cr**; RESTEC, Japan: **145br**; Ronald Royer: **21br**; Simon Fraser: **75l**; Sinclair Stammers: **38t, 59l, 95cr, 100**; Soames Summerhays: **137c**; Worldsat International Inc.: **154b**

Southampton Oceanography Centre, Empress Dock, Southampton: **29br, 39tc, 98**

Sylvia Cordaiy Photo Library: John Farmer: **169t**; Vaughn Evans: **109l**

The Image Bank: **138bl, 147bl, 153b**

The Natural History Museum, London: **44t, 54tr, 54c, 59tr, 63tl, 68t, 80bl, 97cr, 108br, 115tl**

Tony Stone Images: **23bl, 33br, 34, 71b, 71cr, 116, 129, 131, 133br, 137tr, 138br, 141bl, 143tl, 143tr, 143br, 155, 161br, 163br**

Werner Forman Archive, London: **165tl**

Design and Layout:
Simon Burrough
Ralph Orme

Editorial:
Elizabeth Wyse

Production:
Andrea Fairbrass
Barry Haslam
Ralph Orme

Maps Compiled and Produced by:
James East
Andrea Fairbrass
Peter Gamble
Elsa Gibert
Elizabeth Hudson
Isabelle Lewis
David McCutcheon
Kevin Panton
Peter Smith
Malcolm Swanston
Nicholas Whetton

Illustrations:
Julian Baker
Peter Massey
Ralph Orme
Steve Roberts

Indexing:
Jean Cox
Barry Haslam

Picture Research:
Charlotte Taylor

Typesetting:
Jeanne Radford
Charlotte Taylor

Colour Separations:
Central Systems, Nottingham